ATROPOS PRESS
new york • dresden

Deleuze
History and Science

Manuel DeLanda

Think Media EGS Series is supported by the European Graduate School

ATROPOS PRESS
New York • Dresden

151 First Avenue # 14, New York, N.Y. 10003

cover design: Hannes Charen

ISBN 978-0-9827067-1-8

CONTENTS

Acknowledgements.   2

Assemblage Theory and Human History.   3

Materialism and Politics.   29

Assemblage Theory and Linguistic Evolution.   51

Metallic Assemblages.   67

Materialist Metaphysics.   81

Intensive and Extensive Cartography.   115

Deleuze in Phase Space.   141

ACKNOWLEDGEMENTS.

Some of the essays that make up this book are published here for the first time, but some have appeared in other publications in modified form. The publishers acknowledge that some material has been previously published in the following collections:

Deleuzian Social Ontology and Assemblage Theory. In Deleuze and the Social. Edited by Martin Fuglsang and Bent Meier Sørensen. (Edinburgh: Edinburgh University Press, 2006.)

Deleuze, Materialism, and Politics. In Deleuze and Politics. Edited by Ian Buchanan and Nicholas Thoburn. (Edinburgh: Edinburgh University Press, 2008.)

Molar Entities and Molecular Populations in History. In Deleuze and History. Edited by Jeffrey Bell and Claire Colebrook. (Edinburgh: Edinburgh University Press, 2009.)

Deleuze in Phase Space. In Virtual Mathematics. Edited by Simon Duffy. (Manchester: Clinamen Press, 2006.)

Assemblages and Human History.

We no longer believe in a primordial totality that once existed, or in a final totality that awaits us at some future date. We no longer believe in the dull gray outlines of a dreary, colorless dialectic of evolution, aimed at forming a harmonious whole out of heterogeneous bits by rounding off their rough edges. We believe only in totalities that are peripheral. And if we discover such a totality alongside various separate parts, it is a whole of these particular parts but does not totalize them; it is a unity of all those particular parts but does not unify them; rather it is added to them as a new part fabricated separately.

Gilles Deleuze and Felix Guattari. The Anti-Oedipus. [1]

A crucial question confronting any serious attempt to think about human history is the nature of the historical actors that are considered legitimate in a given philosophy. One can, of course, include only human beings as actors, either as rational decision-makers (as in micro-economics) or as phenomenological subjects (as in micro-sociology). But if we wish to go beyond this we need a proper conceptualization of social wholes. The very first step in this task is to devise a means to block micro-reductionism, a step usually achieved by the concept of *emergent properties,* properties of a whole that are not present in its parts: if a given social whole has properties that emerge from the interactions between its parts, its reduction to a mere aggregate of many rational decision makers or many phenomenological experiences is effectively blocked. But this leaves open the possibility of macro-reductionism, as when one rejects the rational actors of micro-economics in favor of society as a whole, a society that fully determines the nature of its members. Blocking macro-reductionism demands a second concept, the concept of *relations of exteriority* between parts. Unlike wholes in which "being part of this whole" is a defining characteristic of the parts, that is, wholes in which the parts cannot subsist independently of the relations they have with each other (relations of interiority) we need to conceive of emergent wholes in which the parts retain a relative autonomy, so that they can be detached from one whole and plugged into another one entering into new interactions.

With these two concepts we can define social wholes, like interpersonal networks or institutional organizations, that cannot be reduced to the persons that compose them, and that, at the same time, do not reduce those persons to the whole, fusing them into a totality in which their individuality is lost. Take for example the tightly-knit communities that inhabit small towns or ethnic neighborhoods in large cities. In these communities an important emergent property is the degree to which their members are linked together. One way of examining this property is to study networks of relations, counting the number of direct and indirect links per person, and studying their connectivity. A crucial property of these networks is their *density*, an emergent property that may be roughly defined by the degree to which the friends of the friends of any given member (that is, his or her indirect links) know the indirect links of others. Or to put it still more simply, by the degree to which everyone knows everyone else. In a dense network word of mouth travels fast, particularly when the content of the gossip is the violation of a local norm: an unreciprocated favor, an unpaid bet, an unfulfilled promise. This implies that the community as a whole can act as a device for the storage of personal reputations and, via simple behavioral punishments like ridicule or ostracism, as an enforcement mechanism.

The property of density, and the capacity to store reputations and enforce norms, are non-reducible properties and capacities of the community as a whole, but neither involves thinking of it as a seamless totality in which the members' personal identity is created by the community. A similar point applies to institutional organizations. Many organizations are characterized by the possession of an authority structure in which rights and obligations are distributed asymmetrically in a hierarchical way. But the exercise of authority must be backed by *legitimacy* if enforcement costs are to be kept within bounds. Legitimacy is an emergent property of the entire organization even if it depends for its existence on personal beliefs about its source: a legitimizing tradition, a set of written regulations, or even for small organizations, the charisma of a leader. The degree to which legitimate authority is irreducible to persons can, of course, vary from case to case. In particular, the more organizational resources are linked to an office or role (as opposed to the incumbent of that role) the more irreducible legitimacy is. Nevertheless, and however centralized and despotic an organization may be, its members remain ultimately separable from it, their actual degree of autonomy depending on contingent factors about social mobility and the existence of opportunities outside the organization.

It is this type of social whole produced by relations of exteriority, wholes that do not totalize their parts, that the opening quote refers to. But that quote also mentions another important characteristic: that the wholes are peripheral or exist alongside their parts. What exactly does this mean? It is not a spatial reference, as if communities or organizations existed nearby or to one side of the persons that compose them. Deleuze and Guattari may simply intend to say that the properties of the whole are not transcendent (existing on a supplementary dimension above its parts) but immanent. But it may also be an ontological or metaphysical remark: communities or organizations, to stick to these examples, are as historically individuated as the persons that compose them. While it is true that the term "individual" has come to refer to persons (or organisms in the case of animals and plants) it is perfectly coherent to speak of individual communities, individual organizations, individual cities, or individual nation states.

In this extended sense the term "individual" has no preferential affinity for a particular scale (persons or organisms) and refers to any entity that is *singular and unique*. Unlike philosophical approaches that make a strong ontological distinction between levels of existence (such as genus, species, organism) here all entities must be thought of as existing at the same ontological level differing only in scale. The human species, for example, is every bit a historical individual as the organisms that compose it. Like them, it has a date of birth (the event of speciation) and, at least potentially, a date of death (the event of extinction). In other words, the human species as a whole exists "alongside" the human organisms that compose it, alongside them in an ontological plane populated only by historically individuated entities.

Historical explanations are inevitably shaped by the ontological presuppositions of the historians who frame them. Historians may be roughly divided into two groups along the lines suggested in the opening paragraph, that is, depending on which of the terms of the following binary oppositions they favor: "the individual versus society", "agency versus structure", "choice versus order". Taking the side of the first terms in these dichotomies yields narratives in which persons, typically "great men", have shaped events, situations, or the outcomes of particular struggles, through their ideas and actions. This does not necessarily imply a disbelief in the existence of society as a whole, only a conception of it that makes it into an epiphenomenon: society is a sum or aggregate of many rational agents or many phenomenological experiences shaped by daily routine. Taking the side of the second terms, on the other hand, yields narratives framed in terms of the transformations that enduring social structures have undergone. The best known example of this is the sequence feudalism-capitalism-socialism. As before, there is no implication here that persons do not exist only that they are a mere epiphenomenon: persons are socialized as they grow up in families and attend schools, and after they have internalized the values of their societies their obedience to traditional regulations and cultural values can be taken for granted.

The late historian Fernand Braudel broke with both of these traditional stances when he set out to study economic history taking as his subject "society as a set of sets." [2] The characters in his narratives include such diverse entities as communities, institutional organizations, cities, and the geographical regions formed by several interacting towns of different sizes. Persons are featured too but not as great men, while larger entities, like kingdoms, empires, world-economies, are treated not as abstract social structures but as concrete historical entities. Speaking of a "sets of sets" is another way of saying that the variety of forms of historical agency (communal agency, organizational agency, urban agency, imperial agency) are related to one another as parts to wholes. Braudel's is a multi-scaled social reality in which each level of scale has its own relative autonomy and hence, its own history. Hence, history ceases to be constituted by a single temporal flow – the short time scale at which personal agency operates or the longer time scales at which social structure changes – and becomes a multiplicity of flows, each with its own variable rates of change.

Braudel's vision can be enriched by replacing his sets, or sets of sets, with the irreducible and decomposable wholes just discussed. Let's illustrate this with a specific example, one that combines Braudel's data with an ontology of individual entities constraining the field of valid historical actors. An entity such as "the Market", for example, would not be an acceptable entity to be incorporated into explanations of historical phenomena because *it is not an individual emergent whole but a reified generality*. But the marketplaces or bazaars that have existed in every urban center since antiquity, and more recently in every European town since the 11th century, are indeed individual entities and can therefore figure as actors in explanations of the rise of Europe, and of the commercial revolution that characterized the early centuries of the second millennium. Equally valid are the regional trading areas that emerged when the towns that housed local marketplaces became linked together by roads and the trade among them reached a threshold of regularity and volume. Regional markets began to play an important economic role in Europe by the 14th century and, as historically constituted wholes composed of local marketplaces, they are valid historical actors. So are the national markets that, starting in England in the 18th century, came into being by stitching together, sometimes forcefully, many provincial trading areas themselves composed of many regional markets. By the 19th century the railroad and the telegraph made the creation of national markets a simpler task and they emerged in places like France, Germany, and the United States, playing an important role in the economic history of these countries. [3]

Other reified generalities, like "the State" should also be replaced. As argued above, in addition to communities a set of interacting persons can give rise to institutional organizations possessing emergent properties like legitimacy. Organizations, in turn, can interact to form a larger whole like a federal government. The latter is a whole in which many organizations are arranged in a hierarchical way with authority operating at different scales: some have a jurisdiction that extends to the entire country; others have authority only within the boundaries of a province or state; and yet others operate within the limits of an urban center and its surrounding region. When it comes to the *implementation* of federal policies this nested set of overlapping jurisdictions can be a powerful obstacle, many policies becoming distorted and weakened as they are implemented at different scales. This problem, however, can become invisible to historians that use the concept of "the State" and view governments as monolithic entities. These two examples illustrate that the distinction between micro and macro should never be made absolute, with individual persons playing the role of micro-entity and society as a whole the role of macro-entity. Rather, micro and macro should be made relative to a particular scale. Compared to the regional trading areas that they compose, local marketplaces are micro while regional markets are macro. But the later are micro relative to provincial markets which are, in turn, micro relative to national markets. Similarly, government organizations with federal jurisdiction can be considered macro relative to those with authority extending only to borders of states or provinces, and these in turn are macro relative to local urban authorities.

Thus, both "the Market" and "the State" can be eliminated from a materialist ontology by a nested set of individual emergent wholes operating at different scales. The expression "operating at different scale", on the other hand, must be used carefully. In particular, it should refer only to *relative scale*, that is, to scale relative to the part-to-whole relation. Given the fact that any emergent whole has always a larger extension than the parts of which it is composed, this relative usage is unproblematic: communities or organizations are always larger than the persons that compose them. But the same is not true if the term "scale" is used in an absolute sense. If instead of comparing a community with its own members, we compared the entire population of persons and the entire population of communities inhabiting a country, for example, we would have to admit that both populations are *coextensive*, that is, that they occupy the same amount of space: the entire national territory. And a similar point applies to the population of institutional organizations. But even if we relativize the concept we may still disagree on the use of the expression "levels of scale" to distinguish social wholes. Why not use, for example, the expression "levels of organization", a phrase used by biologists to characterize the part-to-whole relations between individual cells, individual organs, and individual organisms?. Because this concept carries with it connotations of increased complexity between levels, and in some cases, even teleological implications, as when biological evolution is viewed as involving a drive to greater complexity, from unicellular organisms to multicellular ones. The expression "levels of scale", on the other hand, carries no such connotations: a city is clearly larger than a human being but there is no reason to believe that it possesses a higher degree of complexity, or that any of its component parts is more complex than the human brain.

One final point needs to be clarified: when we say that a set of interacting persons gives rise to a community, or that a set of interacting organizations gives rise to a federal government, this should not be taken to imply a temporal sequence, as if a set of previously disconnected persons or organizations had suddenly began to interact and a whole had abruptly sprouted into being. In a few cases this may indeed be the case, as when people from a variety of war-stricken communities aggregate into a refugee camp and a larger whole emerges from their interactions; or when previously rival industrial organizations aggregate into a cartel forming a larger whole as they interact. But in the majority of cases the component parts come into being when a whole has already constituted itself and has begun to use its own emergent capacities to constrain and enable its parts: most people are born into communities that predate their birth, and most new government agencies are born in the context of an already functioning central government. Nevertheless, the ontological requirement of immanence forces us to conceive of the identity of a community or of a central government as being continuously produced by the day to day interactions between its parts: the emergent properties of a social whole are immanent only to the extent that they would cease to exist if its parts ceased to interact. So we need to include in a materialist ontology not only the processes that historically produce the identity of a given social whole, but also the processes that maintain that identity through time.

Let's pause for a moment to consider how compatible these ideas are with those of Deleuze and Guattari. The first sign of incompatibility is that the expression "the State" occurs throughout their work. But this term is often used as synonymous with "State apparatus", a term that is much less objectionable since it can be taken to refer to the *organizational apparatus* of a given government, that is, to an emergent whole composed of many organizations. A more problematic term, one that is also often used in their historical explanations, is the term "social field" (or less often, "the socius"). This term does indeed refer to "society as a whole" and it is therefore not a valid historical actor in the materialist ontology being sketched here. It is unclear, for example, just what kind of entity this "social field" is supposed to be. Deleuze and Guattari distinguish between different kinds of social wholes: *strata and assemblages*. A State apparatus is classified by them as a stratum. [4] Tightly-knit communities, with their capacity to police their members and punish violations of local norms, would also be a stratum. But an alliance or coalition of several heterogenous communities would be considered to be an assemblage. As Deleuze writes:

What is an assemblage? It is a multiplicity which is made up of heterogeneous terms and which establishes liaisons, relations between them, across ages, sexes and reigns – different natures. Thus the assemblage's only unity is that of a co-functioning: it is a symbiosis, a 'sympathy'. It is never filiations which are important, but alliances, alloys; these are not successions, lines of descent, but contagions, epidemics, the wind. [5]

So we face the problem of whether to treat the "social field" as a stratum or as an assemblage. A different but related problem is that distinguishing between different *kinds* of wholes (strata in general, assemblages in general) may open the back door for reified generalities to infiltrate a materialist ontology. To avoid this danger we can use a single term and build into it "control knobs" (or more technically, parameters) that can have different settings at different times: for some settings the social whole would be a stratum, for other settings an assemblage. The term "parameter" comes from scientific models of physical processes. Whereas variables specify the different ways in which an object being studied is free to change (its "degrees of freedom") parameters specify the environmental factors that affect the object. Temperature can be a variable, the internal temperature of a body of water, for example, as well as a parameter quantifying the degree of temperature of the water's surroundings. Parameters are normally kept constant in a laboratory to study an object under repeatable circumstances, but they may also be varied causing drastic changes in the object under study: while for many values of a parameter like temperature only a quantitative change will be produced, at critical points a body of water will spontaneously change *qualitatively*, abruptly transforming from a liquid to a solid form, or from a liquid to a gas form.

If we parametrized a single concept, then strata and assemblages would cease to be kinds and become *phases*, like the solid and fluid phases of matter. Unlike mutually exclusive binary categories, phases can be transformed into one another, and even coexist as mixtures, like a gel that is a mixture of the solid and liquid phases of different materials. Deleuze and Guattari routinely establish oppositions between kinds (trees and rhizomes, striated and smooth spaces) only to backtrack later as they discuss the ways in which one kind can be transformed into another, or form hybrid mixtures. Thus, the strategy I will follow here will be to keep a single term, the term "assemblage", and parametrize it to allow it to exhibit qualitatively different phases. While we could, of course, parametrize the term "stratum", the first choice is better because the original French term, "agencement", has quite distinct connotations. Thus, we can use the English term "assemblage" to denote the parametrized concept and revert to the French term whenever we need to refer to the original concept. Before discussing the nature of the parameters let's summarize what has been said about assemblages so far:

1) All assemblages have a fully contingent historical identity, and each of them is therefore an *individual entity:* an individual person, an individual community, an individual organization, an individual city. Because the ontological status of all assemblages is the same, entities operating at different scales can directly interact with one another, individual to individual, a possibility that does not exists in a hierarchical ontology, like that composed of genera, species, and individuals.

2) At any level of scale we are always dealing with *populations* of interacting entities (populations of persons, pluralities of communities, multiplicities of organizations, collectivities of urban centers) and it is from the interactions within these populations that larger assemblages emerge as a *statistical result*, or as collective unintended consequences of intentional action. In a given population some entities may get caught into larger "molar" wholes, while other may remain free, composing a "molecular" collectivity. This means that a whole at a given scale is composed not only of molar entities at the immediately lower scale but also of smaller molecular parts.

3) Once a larger scale assemblage is in place it immediately starts acting as a source of limitations and resources for its components. In other words, even though the arrow of causality in this scheme is bottom-up, it also has a top-down aspect: an assemblage both constrains and enables its parts. The upward causality is necessary to make emergent properties immanent: an assemblage's properties may be irreducible to its parts but that does not make them transcendent, since they would cease to exist if the parts stopped interacting with one another. The downward causality is needed to account for the fact that most social assemblages are composed of parts that come into existence after the whole has emerged. Most of the buildings or neighborhoods that compose a modern city, for example, were not only created after the urban center's own birth, but their defining properties were constrained by the city's zoning laws, and their creation made possible by the city's wealth.

Let's now parametrize the concept of assemblage. The first parameter quantifies the *degree of territorialization and deterritorialization* of an assemblage. Territorialization refers not only to the determination of the spatial boundaries of a whole – as in the territory of a community, city, or nation state – but also to the degree to which an assemblage's component parts are drawn from a homogenous repertoire, or the degree to which an assemblage homogenizes its own components. As mentioned before, the members of a densely connected community are constrained by the capacity of the community to store reputations and enforce local norms, a constraint that may result in a reduction of personal differences and in an increased degree of conformity. When two or more communities engage in ethnic or religious conflict, for example, not only the geographical boundaries of their neighborhoods or small towns will be policed more intensely, so will the behavior of their members as the distinction between "us" and "them" sharpens: any small deviation from the local norms will now be observed and punished and the homogenization of behavior will increase. Conflict, in other words, tends to increase the degree of territorialization of communities, a fact that may be captured conceptually by a changing the setting of this parameter.

The second parameter quantifies an assemblage's *degree of coding and decoding*. Coding refers to the role played by language in fixing the identity of a social whole. In institutional organizations, for example, the legitimacy of an authority structure is in most cases related to linguistically coded rituals and regulations: in organizations in which authority is based on tradition, these will tend to be legitimizing narratives contained in some sacred text, while in those governed by a rational-legal form of authority they will be written rules, standard procedures, and most importantly, a constitutional charter defining its rights and obligations. While all individual organizations are coded in this sense, a state apparatus performs coding operations that affect an entire territory and all the communities and organizations that inhabit it. The more despotic or totalitarian a state apparatus the more everything becomes coded: dress, food, manners, property, trade. Because many archaic states allowed the communities over which they ruled to keep their own social codes, superimposing on them a dominant code, Deleuze and Guattari refer to this operation as "overcoding". [6]

Armed with this parametrized concept we can give a more detailed treatment of the different levels of scale at which social entities operate. We can assume that the smallest scale is that of persons, but only as long as the subjectivity of each person is itself conceived as emerging from the interactions between sub-personal components. From the philosopher David Hume, Deleuze derives a conception of the subject or person as an entity emerging from the interactions of a heterogeneous population of sense impressions, and of low-intensity replicas of those impressions (ideas). These sub-personal components are assembled through the habitual application of certain operators to the ideas. More specifically, a subject crystallizes in the mind through the habitual grouping of ideas via relations of contiguity; their habitual comparison through relations of resemblance; and the habitual perception of constant conjunction of cause and effect that allows one idea (that of the cause) to always evoke another (the effect). Perceived contiguity, causality, and resemblance, as relations of exteriority, constitute the three principles of association that transform a mind into a subject. [7]

Deleuze never gave a full assemblage analysis of subjectivity, but it is possible to derive one from his work on Hume. The sub-personal expressive components of the assemblage would comprise both those that are non-linguistic (sense impressions of varying vividness) and those dependent on language, such as beliefs considered as attitudes towards the meaning of declarative sentences (propositions). Material components would include the routine mental labor performed to assemble ideas into a whole, as well as the biological machinery of sensory organs needed for the production of impressions. Habit itself would constitute the main process of territorialization, that is, the process that gives a subject its defining boundaries and maintains those boundaries through time. Habit performs a *synthesis of the present and the past* in view of a possible future. [8] This yields a determinate duration for the lived present of the subject, a fusion of immediately past and present moments, and generates a sense of anticipation, so that habitual repetition of an action can be counted on to yield similar results in the future. A process of deterritorialization, on the other hand, would be any process that takes the subject back to the state it had prior to the creation of fixed associations between ideas, that is, the state in which ideas are connected as in a *delirium*. The onset of madness, high fever, intoxication, sensory deprivation, psychedelic drugs, and a variety of other processes, can all cause a loss or destabilization of subjective identity.

Personal identity, on the other hand, has not only a private aspect but also a public one, the *public persona* that we present to others when interacting with them in a variety of social encounters. Some of these social encounters, like conversations, are sufficiently ritualized that they themselves may be treated as assemblages. The author who has done the most valuable research on conversations is without doubt the sociologist Erving Goffman who defines the subject matter of this research as:

...the class of events which occurs during co-presence and by virtue of co-presence. The ultimate behavioral material are the glances, gestures, positionings, and verbal statements that people continuously feed into the situation, whether intended or not. These are the *external signs of orientation and involvement* – states of mind and body not ordinarily examined with respect to their social organization. [9]

The emphasis on the external signs exchanged during social encounters makes this research ripe for a treatment in terms of emergent wholes in which components are joined by relations of exteriority. While the most obvious expressive component of this assemblage may be the flow of words itself, there is another one which is not always dependent on language. Every participant in a conversation is expressing his or her public identity through every facial gesture, posture, dress, choice of subject matter, the deployment of (or failure to deploy) poise and tact, and so on. These and other components express in a non-linguistic way the image that every participant wants to project to others. The expression of these claims to a public persona must be done carefully: one must choose an image that cannot be easily discredited by others. Any conversation will then be filled with objective opportunities to express favorable information about oneself, as well as objective risks to unwittingly express unfavorable facts. The material components of the assemblage are more straightforward, consisting both of the physical bodies assembled in space, close enough to hear each other and correctly oriented towards one another, as well as the attention needed to keep the conversation going and the labor involved in repairing breaches of etiquette or recovering from embarrassing events. [10] Some technological inventions, such as the telephone, can change the requirement of co-presence, eliminating some of the material components (spatial proximity) but adding others: the technological device itself, as well as the infrastructure needed to link many such devices.

Processes of territorialization giving a conversation well-defined borders in space and time are exemplified by behavior guided by conventions. As assemblages conversations have a temporal structure, in which ways of initiating and terminating an encounter, as well as taking turns during the encounter, are normatively enforced by the participants. The spatial boundaries of these units are clearly defined partly because of the physical requirement of co-presence and because the participants themselves ratify each other as legitimate interactors excluding nearby persons from intruding into the conversation. [11] Embarrassment, damaging as it is to the public personas projected during the encounter, may be viewed as the main destabilizing factor, by taking attention away from the conversation and focusing it on the embarrassed participant. Goffman discusses critical points of embarrassment after which regaining composure becomes impossible to achieve, embarrassment is transmitted to all participants, and the conversation falls apart. [12] But other critical events may take place that transform a conversation into a heated discussion, or an intense argument into a fist fight. These should also be considered deterritorializing factors, as should technological inventions that allow the conversation to take place at a distance blurring its spatial boundaries.

When many conversations among the same groups of participants, or among different but overlapping groups, have taken place a new social entity may emerge: an interpersonal network. This may be a network of friends or professional colleagues living in different places, or the tightly-knit communities that have already been discussed. To analyze this larger assemblage we can use the resources offered by network theory, the only part of theoretical sociology which has been successfully formalized. In the theory of networks the recurring patterns of links between nodes are often more important than the defining properties of the nodes themselves, a fact that orients the theory towards relations of exteriority. The links in a network may be characterized in a variety of ways: by their presence or absence, the absences indicating the borders separating one network from another, or defining a clique within a given network; by their strength, that is, by the frequency of interaction among the persons occupying the nodes, as well as by the emotional content of the relation; and by their reciprocity, that is, by the mutuality of obligations entailed by the link. As argued above, one of the most important properties of an interpersonal network is its density, a measure of the degree of connectivity among its indirect links. [13]

The links in a network must be constantly maintained and the labor involved constitutes one of the material components. This labor goes beyond the task of staying in touch with others via frequent conversations. It may also involve listening to problems and giving advice in difficult situations, as well as a providing a variety of forms of physical help, such as taking care of other people's children. In many communities there exists a division of labor when it comes to the maintenance of relations, with women performing a disproportionate amount of it, particularly those who, by obligation or choice, are involved in full time domestic activities. [14] A variety of expressions of solidarity and trust emerging from, and then shaping, interactions, are a crucial component of these assemblages These range from routine acts like having dinner together or going to church, to the sharing of adversity and the displayed willingness to make sacrifices for the community as a whole. [15] Expressions of solidarity may, of course, involve language, but in this case actions speak louder than words.

As in the case of conversations, the value of the territorialization parameter is closely related to physical proximity. Much as conversations, in the absence of technology, involve face to face interaction, communities structured by dense networks have historically tended to inhabit the same small town, or the same suburb or ethnic neighborhood in a large city. These bounded geographical areas are literally a community's territory and they may be marked, and distinguished from others, by special expressive signs. Deterritorializing processes include any factor that decreases density, promotes geographical dispersion, or eliminates some of the rituals that, like churchgoing, are key to the maintenance of traditional solidarity. Social mobility and secularization are among these processes. The former weakens links by making people less interdependent, by increasing geographical mobility, and by promoting a greater acceptance of difference through less local and more cosmopolitan attitudes. For the same reason, the resulting deterritorialized networks require their members to be more active in the maintenance of links and to invent new forms of communal participation, given that connections will tend to be wider and weaker and that ready-made rituals for the expression of solidarity may not be available. [16] The same kind of resourcefulness in the means to maintain linkages may be needed in interpersonal networks deterritorialized by technology. For example, in the early "virtual communities" that emerged in the internet (such as the Well) the members were aware of the loss that a lack of co-presence involved and special meetings or parties were regularly scheduled to compensate for this. [17]

While in a friendship network a particular node may become dominant by being more highly connected, directly and indirectly, to other nodes, this centrality or popularity rarely gives the person occupying that position the capacity to issue commands to those located in less centrally located nodes. This capacity implies the existence of an authority structure, and this, in turn, means that we are dealing with a different assemblage: an institutional organization. Organizations come in a wide range of scales, with nuclear families at the low end and government bureaucracies and commercial, industrial or financial corporations at the other end. A modern hierarchical organization may be studied as an assemblage given that the relations between its components are relations of exteriority, that is, what holds the whole together are relatively impermanent contractual relations through which some persons transfer rights of control over a subset of their actions to other persons. This voluntary submission breaks the symmetry of the relations among persons in an interpersonal network where a high degree of reciprocity is common. [18]

There is a variety of forms of authority. In small organizations, like religious sects, the charisma of a leader may be enough to legitimize commands but as soon as the number of members increases past a certain threshold, formal authority becomes necessary, justified by a tradition, as in organized religion, or by actual problem-solving performance, as in the case of bureaucracies. [19] In all organizations the automatic obedience to commands on a day to day basis constitutes a powerful expression of legitimacy. For the same reason any act of disobedience, particularly when it goes unpunished, threatens this expression and may damage the morale of those who do obey. Hence, the expressive role of some forms of punishment designed to make an example of transgressors. Punishment, on the other hand, also has a physical aspect, and this points to the material components of the assemblage, related not so much to practices of legitimization as to *practices of enforcement.* In charismatic and traditional organization these practices may involve torture, mutilation, confinement, exile. But in modern bureaucracies, as well as in many other members of the population of organizations (prisons, hospitals, factories, schools, barracks) enforcement uses subtler but perhaps more efficient means: a specific use of space, in which dangerous groupings are broken up and individual persons are assigned a relatively fixed place; systematic forms of inspection and monitoring of activity, a practice that shapes and is shaped by the analytical use of space; and finally, a constant use of logistical writing, like the careful keeping of medical or school records, to permanently store the product of monitoring practices. [20]

As with interpersonal networks, territoriality in the case of organizations has a strong spatial aspect. Most organizations possess physical premises within which they carry on their activities and which, in some cases, define the extent of their jurisdiction. This territory is defined both formally, by the legitimate jurisdictional area, as well as materially, by the area in which authority can actually be enforced. But just as in interpersonal networks, processes of territorialization go beyond the strictly spatial. The routinization of everyday activities, in the form of the repetition of rituals or the systematic performance of regulated activities, stabilizes the identity of organizations and gives them a way to reproduce themselves, as when a commercial organization opens up a new branch and sends part of its staff to bring with them the institutional memory (the day to day routines) of the parent company. Technological innovation, on the other hand, can destabilize this identity, deterritorializing an organization, and opening the assemblage to change. Transportation and communication technologies, for example, can have deterritorializing effects on organizations similar to those on face to face interaction, allowing organizations to break away from the limitations of spatial location. The modern bureaucratic form of authority may have emerged in part thanks to the precision with which the dispersed activities of many branches of an organization could be coordinated via the railroads and the telegraph. [21] And a similar point can be made about the transformation that large commercial or industrial corporations underwent in the nineteenth century, as they became nationwide corporations, as well as in the twentieth century when they became international.

Individual organizations may form larger social entities, such as the supplier and distribution networks linked to large industrial firms, or the already mentioned hierarchies of governmental agencies operating within smaller or larger jurisdictions depending on their rank. Let's skip this important layer to describe the largest scales, such as cities or nation-states. Neither urban centers nor territorial states should be confused with the organizations that make up their government, even if the jurisdictional boundaries of the latter coincide with the geographical boundaries of the former. Cities and nation-states must be viewed as physical locales in which a variety of differently scaled social agents carry on their day to day activities. A city, for example, possesses not only a physical infrastructure and a given geographical setting, but it also houses a diverse population of persons; a population of interpersonal networks, some dense and well localized, others dispersed and shared with other cities; a population of organizations of different sizes and functions, some of which make up larger entities such as industries or sectors. A city assembles the activities of these populations in a concrete physical locale. And similarly for territorial states, from empires and kingdoms to nation-states.

Cities possess a variety of material and expressive components. On the material side, we must list for each neighborhood the different buildings in which the daily activities and rituals of the residents are performed and staged (the pub and the church, the shops, the houses, and the local square) as well as the streets connecting these places. In the nineteenth century new material components were added, water and sewage pipes, conduits for the gas that powered early street lighting, and later on electricity cables and telephone wires. Some of these components simply add up to a larger whole but citywide systems of mechanical transportation and communication can form very complex networks with properties of their own, some of which affect the material form of an urban center and its surroundings. A good example is locomotives (and their rail networks) which possess such a large mass and are so hard to stop and accelerate again, that they determine an interval of two or three miles between stops. This, in turn, can influence the spatial distribution of the suburbs which grow around train stations, giving them their characteristic bead-like shape. [22]

On the expressive side, a good example is a city's skyline, that is, the silhouette cut against the sky by the mass of its buildings and the decorated tops of its churches and public buildings. For centuries these skylines were the first image visitors saw as they approached a city, a recognizable expression of a town's identity, an effect lost later on as suburbs and industrial hinterlands blurred city boundaries. In some cases, the physical skyline of a town is simply a sum of its parts but the rhythmic repetition of architectural motifs – minarets, domes and spires, belfries and steeples – and the counterpoint these motifs create with the surrounding landscape, may produce emergent expressive effects. [23] In the twentieth century skyscrapers and other signature buildings were added to the skyline as a means to make it unique and instantly recognizable, a clear sign that the expressivity of skylines had become the object of deliberate planning.

A variety of territorializing and deterritorializing processes may affect the state of a city's boundaries, making them either more permeable or more rigid, and affecting the sense of geographical identity of its inhabitants. Two extreme forms of these boundaries stand out in Western history. In ancient Greek towns a large part of the population lived in summer months in their rural homes. This double residence and the lack of clearly-defined city boundaries affected their sense of urban identity, as shown by the fact that a town's residents congregated into neighborhoods by their rural place of origin, that is, they maintained their original geographical loyalties. [24] European medieval towns, on the other hand, were surrounded by stone walls, giving not only a definite spatial boundary to the jurisdiction of a town's government, but also a very clear sense of geographical identity to its inhabitants. As the historian Fernand Braudel puts it, these highly territorialized cities "were the West's first focus of patriotism – and the patriotism they inspired was long to be more coherent and much more conscious than the territorial kind, which emerged only slowly in the first states." [25] The development of suburbs and industrial hinterlands, starting in the nineteenth century, blurred the boundaries of urban centers with clear deterritorializing effects. For a while cities managed to hang on to their old identities by retaining their center (which became home for train stations and later on, large department stores) but the further extension of suburbs after World War II and the differentiation of their land uses (retail, wholesale, manufacturing, office space) recreated the complex combinations that used to characterize the old city's center. This process, in effect, created brand new centers in the suburban band deterritorializing the identity of cities. [26]

But centuries before residential suburbs replaced city walls another process was militating against the strong identity of urban centers: a loss of autonomy relative to the emerging territorial states. Once cities were absorbed, mostly through military force, the local patriotism of their citizens was largely diminished. In some areas of Europe strong urban identities were obstacles to the creation of nationwide loyalties. For this reason, the first European territorial states (France, England, Spain) were born in those areas which had remained poorly urbanized as Europe emerged from the shadow of the collapse of the Roman Empire. The regions that witnessed an intense urbanization between the years 1000 and 1300 A.D. (northern Italy, northern Germany, Flanders, and the Netherlands) delayed the formation of larger territorial assemblages for centuries. But between the year 1494, when a French army invaded the Italian city-states for the first time, and 1648, the end of the Thirty Years War, most autonomous cities were brought under control. Indeed, the peace treaty that ended that long war, the treaty of Westphalia, is considered the event that gave birth to international law, that is, the legal system in which territorial states were explicitly recognized as legal actors through the concept of "sovereignty". [27]

As assemblages, territorial states posses a variety of material components. These range from the natural resources contained within their frontiers (mineral deposits like coal, oil, precious metals, agricultural land of varying fertility) to their human populations (a potential source of tax payers and of army and navy recruits). The frontiers (and natural boundaries) defining these assemblages play a material role in relation to other such large entities. That is, each kingdom, empire, or nation-state has a given geostrategic position relative to other territorial entities with which it shares frontiers, as well as material advantages deriving from some natural boundaries such as coastlines which may give it access to important sea routes. After the treaty of Westphalia was signed, future wars tended to involve several national actors. This implies, as the historian Paul Kennedy has argued, that geography affected the fate of a nation not merely through

... such elements as a country's climate, raw materials, fertility of agriculture, and access to trade routes – important though they all were to its overall prosperity – but rather [via] the critical issue of strategical *location* during these multilateral wars. Was a particular nation able to concentrate its energies upon one front, or did it have to fight on several? Did it share common borders with weak states, or powerful ones? Was it chiefly a land power, a sea power, or a hybrid, and what advantages and disadvantages did that bring? Could it easily pull out of a great war in Central Europe if it wished to? Could it secure additional resources from overseas?. [28]

There is also a wide range of expressive components of these larger assemblages, from the natural expressivity of their landscapes to the ways in which they express their military might and political sovereignty. The hierarchies of government organizations operating at a national, provincial, and local scales, played a key role in determining how nationalist allegiances would be expressed in nation-states through flags and anthems, parades and celebrations. The cities that became national capitals also played an important expressive role, the best example of which is the style of urban design that became fashionable in Europe after the Thirty Years War. This style, referred to as the "Grand Manner", transformed the new capitals into Baroque displays of the power of their centralized governments: wide avenues were built and lined with trees; sweeping vistas were created, framed by long rows of uniform facades and punctuated by visual markers, such as obelisks, triumphal arches, or statues; and all the different design elements, including the existing or modified topography, were joined in ambitious, overall geometric patterns. [29]

National capitals also played a territorializing role, homogenizing and exporting to the provinces a variety of cultural materials, from a standard language and currency, to legal codes, and medical and educational systems. Territorialization also had a directly spatial manifestation: the controllability of the movement of immigrants, goods, money and, more importantly, foreign troops, across a nation's borders. While the peace treaty of Westphalia gave frontiers a legitimate legal status, the decades that followed its signing witnessed the most intense effort to rigidify these legal borders through the systematic construction of fortress towns, perimeter walls and citadels. In the hands of the brilliant military engineer Sebastian Le Prestre de Vauban, for example, France's borders became nearly impregnable, maintaining their defensive value until the French Revolution. Vauban built double rows of fortresses in the northern and southeastern frontiers, so systematically related to each other that one "would be within earshot of French fortress guns all the way from the Swiss border to the Channel". [30]

The main deterritorializing processes were those that affected the integrity of these borders. These could be spatial processes such as the secession of a province, or the loss of a piece of territory to another country. But they could also be border-defying economic processes. As the frontiers of territorial states were becoming solidified after the Thirty Years War, some maritime cities which had resisted integration were creating commercial and financial networks that were truly international. Such a maritime city was Amsterdam, the seventeenth-century core of what is today called a *world-economy*: a large geographical area displaying a high degree of economic coherence as well as an international division of labor. [28] A world-economy, in fact, had existed in the West since the fourteenth century, with Venice as its core, but when it acquired global proportions in the seventeenth it became a powerful deterritorializing process for nation-states, governing economic flows that, to this day, easily cross political frontiers.

This admittedly simplified description of *society as an assemblage of assemblages* should serve as a reminder of how misleading it is to view human history as comprising a single temporal flow, whether the flow of multiple personal biographies or the one made of the slow glacial movements that affect a society's structure. Indeed, given that even at the largest scale that social assemblages can take (territorial states, world-economies) we never reach a point at which we may coherently talk of "society as a whole", the very term "society" should be regarded as a mere convenient expression. That is, the term should not be considered to have a referent the existence of which we are committed to assert. This is perhaps the way to treat terms like "the social field" or "the socius", terms that constantly appear in the original version of assemblage theory: convenient general expressions that can be replaced when necessary by a description of a concrete assemblage. Only then will philosophy catch up with the groundbreaking research of materialist historians like Fernand Braudel.

REFERENCES:

1. Gilles Deleuze and Felix Guattari. Anti-Oedipus. (New York: Viking, 1977), p. 42.

2. Fernand Braudel. The Wheels of Commerce. (New York: Harper and Row, 1982.), p. 458-459.

3. Fernand Braudel. The Perspective of the World. (New York: Harper and Row, 1986). p. 280-287.

4. Gilles Deleuze and Felix Guattari. A Thousand Plateaus. (New York: University of Minnesota Press, 1987), p. 352.

5. Gilles Deleuze and Claire Parnet. Dialogues II. (New York: Columbia University Press, 2002), p. 69.

6. Gilles Deleuze and Felix Guattari. A Thousand Plateaus. Op. Cit. p. 448.

7. Gilles Deleuze. Empiricism and Subjectivity. (New York: Columbia University Press, 1991), p. 98-101.

8. Gilles Deleuze. Difference and Repetition. (New York: Columbia University Press, 1994), p. 70-74.

9. Erving Goffman. Interaction Ritual. Essays on Face-to-Face Behavior. (New York: Pantheon Books, 1967), p. 1. (My italics)

10. Ibid. p. 19.

11. Ibid. p. 34.

12. Ibid. p. 103.

13. John Scott. Social Network Analysis. (London: Sage Publications, 2000), p. 70-73.

14. Graham Crow. Social Solidarities. (Buckinham: Open University Press, 2002), p. 52-53.

15. Graham Crow. Ibid. p. 119-120.

16. John Scott. Social Network Analysis. Op. Cit. p. 12.

17. Howard Rheingold. The Virtual Community. Homesteading on the Electronic Frontier. (New York: Harper Perennial, 1994).

18. James S. Coleman. Foundations of Social Theory. (Cambridge, Mass.: Belknap Press, 2000). p. 66.

19. Max Weber. The Theory of Social and Economic Organization. (New York: Free Press of Glencoe, 1964). p. 328-336.

20. Michel Foucault. Discipline and Punish. The Birth of Prison. (New York: Vintage Books, 1979) p. 195-196.

21. Max Weber. The Theory of Social and Economic Organization. Op. Cit. p. 363.

22. James E. Vance Jr. The Continuing City. Urban Morphology in Western Civilization. (Baltimore: The John Hopkins University Press, 1990). p. 373.

23. Spiro Kostoff. The City Shaped. Urban Patterns and Meanings Throughout History. (London: Bulfinch Press, 1991) p. 284-285.

24. James Vance. The Continuing City. Op. Cit. p. 56.

25. Fernand Braudel. The Structures of Everyday Life. (Berkeley: University of California Press, 1992.) p. 512.

26. James Vance. The Continuing City. Op. Cit. p. 502-504.

27. J. Craig Barker. International Law and International Relations. (London: Continuum, 2000). p. 5-8.

28. Paul Kennedy. The Rise and Fall of the Great Powers. Economic Change and Military Conflict from 1500 to 2000. (New York: Random House, 1987). Page 86. (Emphasis in the original).

29. Spiro Kostoff. The City Shaped. Op. Cit. p. 211-215.

30. Christopher Duffy. The Fortress in the Age of Vauban and Frederick the Great. (London: Routledge and Kegan Paul, 1985.) p. 87.

31. Fernand Braudel. The Perspective of the World. Op. Cit. p. 21.

Materialism and Politics.

For most of its history leftist and progressive politics has been securely anchored on a materialist philosophy. The goal of improving the material conditions of workers' daily life, of securing women's rights to control their bodies, of avoiding famines and epidemics among the poor: all of these were worthy goals presupposing the existence of an objective world in which suffering, exploitation, and exclusion needed to be changed by equally objective interventions in reality. To be sure there was room in this materialism for the role of subjective beliefs and desires, including those that tended to obscure the objective interests of those whose lives needed improvement, but these were never allowed to define what reality is. The concept of "ideology" may be inadequate for analyzing those beliefs and desires, but it nevertheless captured the fact that there is a material reality with respect to which those subjective states should be compared.

Then everything changed. Idealism, the ontological stance according to which the world is a product of our minds, went from being a deeply conservative position to become the norm in many academic departments and critical journals: cultural anthropologists came to believe that defending the rights of indigenous people implied adopting linguistic idealism and the epistemological relativism that goes with it; micro-sociologists correctly denounced the concept of a harmonious society espoused by their functionalist predecessors, but only to embrace an idealist phenomenology; and many academic departments, particularly those that attach the label "studies" to their name, completely forgot about material life and concentrated instead on textual hermeneutics. To make things worse this *conservative turn* was concealed under several layers of radical chic, making it appealing to students and even activists pursuing a more progressive agenda.

It would take an entire book to document these claims in the detail that they deserve. In the space of this essay I can give only a single example, but one that perfectly illustrates the perverse nature of the conservative turn. It concerns a book that, on the surface, should have given a big boost to materialist politics: Michel Foucault's "Discipline and Punish". As is well known, in this book Foucault analyses a historical transformation in the means to enforce authority, a transformation that took place in Europe in the seventeenth and eighteenth centuries in institutional organizations like prisons, schools, hospitals, barracks, and factories. Although physical torture and confinement are sadly still very much with us, they were replaced in some parts of the population of organizations by subtler means of enforcement: the spatial partitioning of the architecture, and the analytical distribution of human bodies, to facilitate monitoring and control; the increased systematicity of observation and surveillance; and the continuous recording in writing of every detail about performance and behavior. [1]

Foucault breaks new ground with this book, even relative to his own previous work, by giving equal attention to the discursive and non-discursive practices of those in positions of authority in institutional organizations. A discursive practice is one that, as its name implies, produces a discourse: the discourse of criminology, of pedagogy, of clinical medicine, of scientific management (Taylorism). Discourses were, of course, the subject of Foucault's previous publications so it is not surprising that they are still important in this book. But a new set of practices is now added to those that produce discourse, practices that involve causal interventions on the human body: from torture and mutilation, to subtler varieties of punishment, such as imposed physical exercise. Even the systematic keeping of records, a practice that involves writing and could therefore be considered discursive, is indeed non-discursive: it makes use of a logistical form of writing – keeping track of dosages and visits in hospitals; of daily behavior and performance in schools and barracks; of the content of warehouses and raw materials used in factories – a type of writing that may serve as data for those who develop a discourse, but that does not lend itself to endless hermeneutic rounds as real discourses do.

Despite Foucault's clear distinction a majority of those humanities professors that are interested in his work consider torture, physical confinement, drilling, and monitoring to be *discursive practices*: to them that is the achievement of Foucault, to have shown that many things that seem physical and material are actually linguistic. This bastardization of Foucault must not go unchallenged, and his original distinction must be upheld. To put it in a nutshell: while pairing a certain category of crime, like stealing, with a certain category of punishment, like cutting off the thief's hand, is clearly a discursive practice, the actual act of mutilation is equally clearly a non-discursive one. The reduction of the non-discursive, to think of mutilation as a "deconstruction of the body" as one clueless academic once remarked to me, is a symptom of a deep political conservatism hidden under radical chic.

Coping with the conservative turn in American universities is not the only challenge facing the left today. A more important one is to fix the shortcomings of the forms of materialism that are part of its tradition. When one asserts the mind-independence of the material world a crucial task is to explain the more or less stable identity of the entities that inhabit that world. If this identity is explained by the possession of an atemporal essence then all one has done is to reintroduce idealism through the back door. Thus, a coherent materialism must have as its main tool a concept of *objective synthesis*, that is, of a historical process that produces and maintains those stable identities. In traditional forms of materialism, those associated with Marxism, this concept was borrowed from Hegelian idealism but turned right side up, so to speak. The synthetic process in question was, of course, the negation of the negation, the synthesis of opposites. This concept was thought to apply not only to human affairs, the synthesis of new institutions in the cauldron of social conflict, but to also represent a general approach to the dialectics of nature itself. Unfortunately, an apriori concept of synthesis is bound to fail to capture all the different processes through which identity is generated, even if it is turned on its head.

As part of his rejection of Hegelian dialectics, and of a broader rejection of negation as a fundamental concept, Gilles

Deleuze introduced new ideas with which to conceptualize the temporal synthesis of objective entities. In his work with Felix Guattari, for example, he gave us the concept of a process of double articulation through which geological, biological, and even social strata are formed. The first articulation concerns *the materiality of a stratum*: the selection of the raw materials out of which it will be synthesized (such as carbon, hydrogen, nitrogen, oxygen, and sulfur for biological strata) as well as the process of giving populations of these selected materials some statistical ordering. The second articulation concerns the *expressivity of a stratum*. Although in the heavily linguisticized century in which these ideas were written the term "expression" was synonymous with "linguistic expression", in the theory of double articulation the term refers in the first place to material expressivity, that is, to the color, sound, texture, movement, geometrical form, and other qualities that can make geological or meteorological entities so dramatically expressive. This second articulation is therefore the one that consolidates the ephemeral form created by the first and that produces the final material entity defined by a set of emergent properties that express its identity. In the words of Deleuze and Guattari:

> Each stratum exhibits phenomena of *double articulation* ... This is not at all to say that the strata speak or are language based. Double articulation is so extremely variable that we cannot begin with a general model, only a relatively simple case. The first articulation chooses or deducts, from unstable particle-flows, metastable molecular or quasi-molecular units (*substances*) upon which it imposes a statistical order of connections and successions (*forms*). The second articulation establishes functional, compact, stable substances (*forms*), and constructs the molar compounds in which these structures are simultaneously actualized (*substances*). In a geological stratum, for example, the first articulation is the process of "sedimentation" which deposits units of cyclic sediment according to a statistical order: flysch, with its succession of sandstone and schist. The second articulation is the "folding" that sets up a stable functional structure and effects the passage from sediment to sedimentary rock. [2]

There is, in fact, an error in the example given by Deleuze and Guattari. The synthesis of sedimentary rock proceeds by the sorting out of pebbles of different size and composition, an operation performed by the rivers that transport and deposit the raw materials at the bottom of the ocean. These loose accumulations are then cemented together and transformed into layers of sedimentary rock, that is, of an entity with emergent properties not present in the component pebbles. Then, *at a different scale*, many of these emergent rocks accumulate on top of one another and are then folded by the clash of tectonic plates to produce a new emergent entity: a folded mountain range like the Himalayas or the Rocky Mountains. We will see below that this is not the only place where Deleuze and Guattari fail to make a distinction between strata operating at different scales. But the ease with which the mistake can be corrected shows that the concept of a double articulation is robust against simple errors and, more importantly, capable of multiple variations that accommodate the complexity of actual strata. What really matters is not to confuse the two articulations with the distinction between form and substance, since each articulation operates through form and substance: the first selects only some materials, out of a wider set of possibilities, and gives them a statistical form; the second gives these loosely ordered materials a more stable form and produces a new, larger scale material entity. Deleuze and Guattari use a variety of terms to refer to each of these two articulatory operations. Here I will stick to one pair: the first articulation is called "territorialization" and concerns a *formed materiality,* the second one "coding" and deals with a *material expressivity.*

We can now summarize the idea of a double synthesis this way: all the entities that populate the world come into being through specific temporal processes that affect both their materiality and their (nonlinguistic) expressivity. All identities are, in this sense, historical, as long as the word is used to refer not only to human history but to geological, biological, and even cosmic history. This constitutive historicity implies that objective entities are inherently changeable: they may undergo destabilizing processes affecting their materiality, their expressivity, or both. In other words, they may be subject to processes of *deterritorialization and decoding.* This is important in the context of human politics because it is the possibility of social change that is at stake here, as well as the historicity of all social institutions. Whatever one may think about the old historical and dialectical forms of materialism they at least got that right. Finally, there is the question of the role that language plays in all this. In the theory of double articulation the historical emergence of language is treated in a similar way as that of the genetic code. While before the rise of living creatures all expression was three dimensional – the geometry of a crystal, for example, was what expressed its identity – genes are a one-dimensional form of expression, a linear chain of nucleotides, and this *linearization* allows material expressivity to specialize. As Deleuze and Guattari put it:

> Before, the coding of a stratum was coextensive with that stratum; on the organic stratum, on the other hand, it takes place on an autonomous and independent line that detaches as much as possible from the second and third dimensions. ...The essential thing is *the linearity of the nucleic sequence.* ... It is the crystal's subjugation to three-dimensionality, in other words, its index of territoriality, that makes the structure incapable of formally reproducing and expressing itself; only the accessible surface can reproduce itself, since it is the only deterritorializable part. On the contrary, the detachment of a pure line of expression on the organic stratum makes it possible for the organism to attain a much higher threshold of deterritorialization, gives it a mechanism of reproduction covering all the details of its complex spatial structure, and enables it to put all its interior layers topologically in contact with the exterior, or rather with the polarized limit (hence the special role of the living membrane). [3]

Language emerges in a similar way except that its linearity is now temporal not spatial, involving a more intense deterritorialization that makes it even more independent of its formed materiality. This is what gives language the ability to represent all other strata, to translate "all of the flows, particles, codes, and territorialities of the other strata into a sufficiently deterritorialized system of signs...". [4] And this capacity to represent or translate all other strata is, in turn, what gives language, or more exactly language-based theories, their "imperialist pretensions". In other words, the linguisticization of world-views that took place in the twentieth century after the so-called "linguistic turn", forming the basis for the rejection of materialism and the spread of conservative idealism, can be explained within the theory of double articulation as a result of the unique status of this specialized line of expression. Thus explained, the power of language can be accepted while the conceptual obstacle represented by its illegitimate extension circumvented.

Before discussing materialist politics one conceptual obstacle needs to be removed. Traditionally, the sciences that have the most political relevance have divided themselves along micro-macro lines. For a while classical economics represented the micro side, the rational decision maker, and classical sociology the macro side, society as a whole. Eventually, however, both fields diversified: micro-economics was supplemented by Keynes' macro-economics, dealing with macro-quantities like gross national product, and overall inflation and unemployment rates, while the macro-sociology of Durkheim and Parsons was challenged by phenomenologists in the 1960's and gave rise to several forms of micro-sociology, dealing with the effects of daily routine or the effects of stereotypes in shaping personal experience. But there is something deeply wrong with this treatment of the micro and the macro as absolute scales.

A more adequate approach would be to treat them *as relative to a particular scale*. Persons are micro-entities if one is dealing with the community of which they form a part, but macro-entities if one is studying the sub-personal sensations and feelings, beliefs and desires, from which persons crystallize. Communities are macro-entities in relation to the persons that compose them but they may also become part of a larger whole, as when several of them are linked through alliances to form a social justice movement. In that case, a single community is a micro-entity while the entire coalition is a macro-entity. Persons can also be component parts of institutional organizations, that is, organizations possessing an authority structure. In this case persons operate at the micro-level while the entire organization works at the macro-level. But organizations can become parts of larger wholes, such as an industrial network of economic organizations, or a government hierarchy of federal, state, and local organizations. In this case, an industrial network or a federal government are macro-entities, while their component organizations are micro-entities.

Making the micro-macro distinction relative to a specific scale, or more exactly, to a specific relation between parts and wholes, removes this conceptual obstacle. Using this relativized notion of scale, we can think of the two articulations as operating at the micro and macro levels of scale respectively. Deleuze does distinguish several levels at which territorialization and coding may occur: the individual, the group, and the social field. [5] But these three levels may not give us a sufficiently detailed social ontology. In general, what needs to be excluded from a materialist philosophy are vague, reified general terms like "the Market" or "the State". [6] The term "Society" (or the "social field") is not as problematic, since it can always be replaced by concrete wholes like city-states or nation-states, kingdoms or empires. Let's examine double articulation in more detail from this point of view, starting with institutional organizations like prisons, hospitals, schools, barracks, factories and so on. These are, of course, the "species" of organizations whose mutation during the seventeenth and eighteenth centuries was so thoroughly studied by Foucault. In his book on the subject Deleuze distinguishes the two articulations involved in the production of these social entities this way:

> Strata are historical formations, positivities or empiricities. As 'sedimentary beds' they are made from things and words, from seeing and speaking, from the visible and the sayable, from bands of visibility and fields of sayability, from contents and expressions. We borrow these terms from Hjemslev, but apply them to Foucault in a completely different way, since content is not to be confused here with a signified, nor expression with a signifier. Instead, it involves a new and very rigorous division. The content has both form and substance: for example, the form is prison and the substance is those that are locked up, the prisoners ... The expression also has a form and a substance: for example, the form is penal law and the substance is 'delinquency' in so far as it is the object of statements. Just as penal law as a form of expression defines a field of sayability (the statements of delinquency), so prison as a form of content defines a place of visibility ('panopticism', that is to say, a place where at any moment one can see everything without being seen). [7]

Deleuze is here distinguishing the two articulations roughly along the lines of the non-discursive (territorialization) and the discursive (coding). Non-discursive practices of visual surveillance and monitoring, performed in buildings specifically designed to facilitate their routine execution, sort the raw materials (human bodies) into criminal, medical, or pedagogic categories; and discursive practices, like those of criminologists, doctors, or teachers who produce the categories and the discourses in which they are embedded, consolidate those sorted materials giving prisons, hospitals, and schools a more stable form and identity. The only problem with this formulation is the absolute use of the micro-macro distinction. For Deleuze and Foucault, the visible and the articulable define an "disciplinary age" that is, a historical period defining a whole "society". But as in the case of geological strata, the problem is relatively easy to fix.

The first thing that needs to be done is to think of the two articulations as applying to a *population of institutional organizations*, not to "society as a whole", and to add to the the the prisoners processed by prisons, the students processed by schools, and the patients processed by hospitals, the people that staffs those organizations: not just guards, teachers, doctors, nurses, but the entire administrative staff. These are also material components of organizations and, indeed, also subject to surveillance, even if to a lesser degree. Many other organizations, from bureaucracies to large churches, share this administrative staff, but do not have a separate set of bodies to confine and monitor. What all these organizations do have in common is possession of an *authority structure*. Authority has two aspects: legitimacy and enforcement. Foucault focuses on the latter in an effort to go beyond the problematic of legitimacy. But however important it was for his work to stress enforcement practices, practices of legitimization must also be taken into account. Roughly, it is practices of enforcement – including not only visibilities, that is, surveillance, but also the keeping of biographical records and the disciplining of bodies – that constitute the first articulation, while practices of legitimization perform the second articulation.

If Michel Foucault can be considered the first thinker who correctly conceptualized enforcement practices, Max Weber is certainly the one that gave us the best conceptualization of practices of legitimization. He argued that in an organization in which human activity is subject to *imperative coordination* purely coercive measures and material benefits (e.g. wages) are not sufficient to stabilize authority. In addition, those who obey must believe in the legitimacy of those commands, or more exactly, in *the legitimacy of the claims to authority expressed* by those commands. Since Weber considered legitimacy an important source of voluntary submission to commands he classified types of authority in organizations accordingly. Imperative coordination of social activity can occur, according to this classification, in a continuum of forms defined by three "ideal types" and their mixtures.

One pole of the continuum is defined by the extreme case of a perfectly efficient bureaucracy, in which a complete separation of position or office from the person occupying it has been achieved, and in particular, in which a sharp separation of the incumbent from the resources connected to a position has been effected. [8] In addition, the sphere of the incumbent's competence must be clearly defined by written regulations, some of which specify technical rules the application of which may demand specialized training. The official examinations that test incumbents for these technical capacities further solidify the separation of position and occupant. Finally, the positions or offices must form a clear hierarchical structure in which relations of subordination between positions (not persons) are clearly specified in writing, that is, in a legal constitution. Weber refers to this ideal type as "rational-legal" to capture both the constitutional and technical aspects of its order. In this case, obedience is owed to the impersonal order itself, that is, legitimacy rests on a belief in both the legality and technical competence of claims to authority. [9]

Another ideal pole defining the continuum of authority forms is the "traditional type" in which a clear separation between offices and incumbents does not exist. To begin with, obedience is owed to the person occupying a position of authority justified in terms of traditional rules and ceremonies assumed to be sacred. While custom defines the extent of authority of the chief there is also a sphere of personal prerogative within which the content of legitimate commands is left open and may become quite arbitrary. As Weber says, "In the latter sphere, the chief is free to confer grace on the basis of his personal pleasure or displeasure, his personal likes and dislikes, quite arbitrarily, particularly in return for gifts which often become a source of regular income." [10] Finally, the third ideal pole of the continuum involves imperative coordination in which neither abstract legality nor sacred precedent exist as sources of legitimacy. Routine control of collective action on either basis is specifically repudiated by an individual who is treated by followers as a leader by virtue of personal charisma. In reality, the continuum defined by these three ideal types (a possibility space defined by three singularities) will tend to be populated by organizations displaying a mixture of these characteristics: a bureaucracy led by a charismatic elected official, or a bureaucracy in which written rules that used to be means to an end have become ends in themselves, that is, have become ritualized. [11]

We may say that an institutional organization is territorialized to the extent that the human bodies that compose it have been sorted out into the ranks of a hierarchy. The higher the degree of centralization of decision-making and the sharper the definition of the ranks, the more intensely territorialized the organization may be said to be. The degree of territorialization also increases the more obvert punitive interventions on the human body are. Thus, an organization in which torture and indiscriminate confinement are the main means of enforcing authority is more territorialized than one in which enforcement has become more diffused, relying on less obvert forms like daily drill, inconspicuous monitoring, behind the scenes record keeping. The second articulation involves both the discourses produced in these organizations (whether they are merely legitimizing narratives or formal knowledge used to perfect enforcement practices) as well as the ways in which their practices are coded, from written regulations and rationalized daily routines to ritualized behavior and ceremonial dress. The more these routines and rituals are rigidly specified in writing the more coded the organization may be said to be.

Foucault emphasized the fact that modern organizations had a double origin, that is, that each of the two articulations had a separate historical source. The two articulations converged in the Napoleonic state the foundations of which, as Foucault writes:

> ...were laid out not only by jurists, but also by soldiers, not only counselors of state, but also junior officers, not only the men of the courts, but also the men of the camps. The Roman reference that accompanied this formation certainly bears with it this double index: citizens and legionnaires, law and maneuvers. While jurists or philosophers were seeking in the pact a primal model for the construction or reconstruction of the social body, the soldiers and with them the technicians of discipline were elaborating procedures for the individual and collective coercion of bodies. [12]

If this analysis is correct then it is clear that we must go beyond Deleuze's "visibilities and sayabilities". While this way of framing the problem may be useful for *epistemological purposes* – highlighting the role played by organizations in making visible certain aspects human behavior (task performance, medical symptoms, personal predispositions and liabilities) and allowing their discursive articulation – it is much less useful for political purposes, that is, for the purpose of changing the way in which imperative coordination of human activity is carried out in organizations. In particular, understanding the double historical source of legitimacy and enforcement in the rational-legal form (jurists and soldiers) is crucial for any political undertaking that attempts to bring real change. But above all what is crucial for politics is to situate the analysis at the right level of scale. That is, we should avoid the mistake of thinking that we have discovered the essence of the "disciplinary society", when all we have achieved is figuring out how certain practices of enforcement propagated through a population of organizations in the seventeenth and eighteenth centuries.

Let me give one more example of how to apply this extended version of the double articulation theory to other concrete social entities, such as local communities. Many communities exist in well defined spatial locations, like a small town or an ethnic neighborhood in a large city. Their degree of territorialization can be measured by the density of the connections that define their networks of kin and friendship. An inter-personal network in which everybody knows everybody else has a high degree of density and this gives a community the capacity to sort people into insiders and outsiders, and the insiders into those with good and bad reputations. This sorting operation can be said to constitute the first articulation. The second articulation involves expressions of solidarity. Solidarity may be expressed verbally, in speeches directed to a community, but it is expressed more clearly by actual behavior, such as providing physical help or emotional support when it is needed. On the other hand, language does play a role in the storage of a community's memory, in the form of stories of resistance to authority, or stories about conflict with other communities. [13] Regularly listening to these stories increases internal cohesion and consolidates communal identity, and to that extent the practice of story-telling also performs a second articulation.

The degree of solidarity in a community is clearly important in determining the extent to which it may be mobilized for political purposes. Social justice movements, particularly before the rise of long-distance communication technologies, depended on such internal solidarity to create coalitions of communities. These alliances were crucial from the moment expressions of political dissent were transformed in the eighteenth century from machine breaking, physical attacks on tax collectors, and other forms of direct action, to the very different set of displays characteristic of today's public demonstrations. This is a change in what the historical sociologist Charles Tilly calls *repertoires of contention*: the sets of performances through which collective actors express their claims to political rights. These expressive repertoires changed dramatically during the Industrial Revolution, to include "public meetings, demonstrations, marches, petitions, pamphlets, statements in mass media, posting or wearing of identifying signs, and deliberate adoption of distinctive slogans." [14] Through these means a social justice movement could express that it was *respectable, unified, numerous and committed*, in short that it was a legitimate collective maker of claims in the eyes of both its rivals and the government.

The expression of these properties can, of course, be performed by using language. Publishing a statement about the quantity of supporting members will express numerousness, but to the extent that these verbal statements can be exaggerated assembling a very large crowd in a particular place in town will express numerousness more dramatically. The degree of unity in a coalition can be easily expressed verbally, but it will be expressed more forcefully by concerted action and mutual support. But to whom are these dramatic, forceful, convincing claims being expressed? Since these are expressions of claims to specific rights (the right to collective bargaining, to vote, to assemble) the intended audience is typically the governmental organizations that can grant those rights. As Tilly puts it:

> Claim making becomes political when governments – or more generally, individuals or organizations that control concentrated means of coercion – become parties to the claims, as claimants, objects of claims, or stake holders. When leaders of two ethnic factions compete for recognition as valid interlocutors for their ethnic category, for example, the government to which interlocutors would speak inevitably figure as stake holders. Contention occurs everywhere, but contentious politics involves governments, at least as third parties. [15]

As this example illustrates, giving correct historical explanations of political movements involves a more detailed breakdown of social entities. That is, we need to go beyond the three part distinction between the individual, the group, and the social field. And a similar point applies to the case of political economy. In particular, we have become used to speak of a "capitalist society" or the "capitalist system". These terms used to belong to the left, but since the 1980's they have also been adopted by the right, the only difference being that while one side demonizes them the other side glorifies them. The term "capitalism" has degenerated into a word that is part of a morality tale. Deleuze and Guattari have attempted to breathe new life into the concept by redefining it as an *axiomatic of decoded and deterritorialized flows*. The point of the term "axiomatic" is to create a contrast with the relatively fixed form of coding performed by a state apparatus: fixed codes of behavior and dress for different social classes; fixed laws based on ancient writings; fixed repertoires of technology kept closed by fear of innovation, and so on. An axiomatic is, in the field of logic and mathematics, a small body of self-evident truths from which an infinite number of theorems can be derived. Similarly, the "capitalist system" is here conceived as capable of deriving an infinite number of new entities – technologies, customs, fashions, financial instruments – all of which can be made compatible with the overall system. [16]

There is no doubt that the commercial revolution that swept Europe from the thirteenth century on, and the even more intense industrial revolution that started in the eighteenth, had deterritorializing and decoding effects of all kinds. But the question is: what social entities underwent these deterritorializations and decodings?. From a materialist point of view, only social entities that actually exist can be so affected so the question becomes: is there such a thing as "the capitalist system"? Deleuze and Guattari, for whom the marxist tradition was like their Oedipus, the little territory they did not dare to challenge, would say "yes". But for that very reason they can't be trusted in these matters. So who can we trust?. Those economic historians that are the true experts on the subject and that are not bound by allegiance to a tradition. Fernand Braudel, for example, claims that "We should not be too quick to assume that capitalism embraces the whole of western society, that it accounts for every stitch in the social fabric." [17] And he goes on to say that "if we are prepared to make an unequivocal distinction between the market economy and capitalism ... economic solutions could be found which could extend the area of the market and would put at its disposal the economic advantages so far kept to itself by one dominant group of society." [18]

These are powerful words. But how can anyone dare to suggest that we must unequivocally distinguish capitalism from the market economy.? These two terms are, for both the left and the right, strictly synonymous. But if we discard reified generalities like "society as a whole" and concentrate on populations of commercial, financial, and industrial organizations, then the distinction makes perfect sense. More specifically, what Braudel is arguing here is that there have been two economic dynamics in the West ever since the first commercial revolution: wholesale was never like retail (until the second half of the twentieth century), and large industrial production had nothing to do with small scale industry. In other words, he is redefining the word "capitalism" to mean "big business". Personally I do not think that redefinitions are very useful, particularly with terms like "capitalism" that are so deeply entrenched in our discursive practices. But leaving questions of language aside, what Braudel is arguing here is that in the population of economic organizations we can make a distinction between those that due to their large scale can exercise economic power and those that can't. This locates one of the relevant scales at which deterritorializations and decodings take place.

An industrial firm that generates wealth through economies of scale is deterritorialized in a variety of ways. It most likely has the legal form of a joint stock corporation, that is, an organizational structure in which control of day to day operations has been separated from ownership: managers, who move freely from one corporation to another, exercise control, while ownership is dispersed into many stockholders. This is in stark contrast with small firms ran by an entrepreneur who is both the owner and the one who supplies direction for the firm. Large scale also allows corporations to internalize a variety of economic functions either through vertical integration (buying its suppliers or distributors) or horizontal integration (buying firms in different areas). Internalization, in turn, gives these large firms geographical mobility by making them self-sufficient: they can relocate factories and headquarters to any part of a nation state that offers them lower wages and taxes. Today, of course, this mobility has become global, an even more intense deterritorialization. Small firms, particularly those that exist in networks and depend on the agglomeration of talent in a particular geographical area, lack this mobility.

But Braudel also describes other social entities, operating at different scales, that can also be said to have undergone deterritorializations and decodings: cities. Cities can be classified in different ways but a relevant distinction for present purposes is between landlocked cities acting as regional capitals – and later on, national capitals – from those that are maritime ports and act as gateways to the outside through their participation in international trade. Until the advent of the locomotive, sea transport was much more rapid than its terrestrial counterpart and this, together with day to day contact with the open spaces formed by the seas and oceans, made maritime ports less territorialized than landlocked cities. From early on in the past millennium these deterritorialized cities entered into networks through which everything – goods, money, people, news, contagious diseases – moved at a faster speed. In addition, regional capitals like Paris, Vienna, or Madrid, attracted migrants from throughout the region they dominated, and over a few centuries they slowly distilled a unique regional culture, giving them a well defined identity. In other words, these cities were also highly coded. Maritime gateways like Venice, Genoa, Lisbon, or Amsterdam, on the other hand, never acquired a sharp cultural identity since they mixed and matched elements from the variety of alien cultures with which they came into regular contact, making their identity less coded. [19] According to Braudel, it was these deterritorialized and decoded cities that were the birthplace of capitalism, properly redefined.

Finally, we can observe a variety of deterritorializing and decoding effects at many other scales, from individual persons and individual communities to individual nation states. Communities increase in territorialization with the degree of density of their inter-personal networks. Hence, anything that decreases density will deterritorialize them. One of these density-reducing factors is social mobility, a factor that became more and more important as middle classes increased in numbers, and as forms of movable wealth (money, debt in paper, stocks) increased relative to those that were immovable (land). Affordable long distance transportation and communication technologies also acted as deterritorializing forces. I could add many more examples that both confirm Deleuze and Guattari's hypothesis while at the same time showing how inadequate it is to ascribe those deterritorializations and decodings to the "system as a whole", that is, to society as an axiomatic.

Why are Deleuze and Guattari so deeply committed to this idea? Because as I said they remained until the end of their lives under the spell of the bankrupt political economy of Marx. Thus, they write: "If Marx demonstrated the functioning of capitalism as an axiomatic, it was above all in the famous chapter of the tendency of the rate of profit to fall." [20] The problem with this argument is that this "tendency" is entirely fictitious, its only basis being the marxist theory of value: if wage labor is inherently a mode of surplus extraction, that is, if every bit of profit that an industrial organization makes is ultimately a product of labor, and if machines are merely the coagulated labor of the those workers that put its parts together, then as capitalists replace humans with machines there will necessarily be a fall in profits. But has anyone ever produced evidence that a factory ran by robots does not produce any profits? Of course not. Machines also produce value (and profits) because they are not just a product of labor but much more importantly, of engineering design and science. And industrial organization (like Taylorism) is also a source of value, even if it carries hidden costs like the deskilling of a worker population.

The belief in the tendency of the rate of profit to fall, or rather, on the labor theory of value that underpins it, leads Deleuze and Guattari to deny Braudel's well documented assertion that maritime metropolises in the thirteenth century were the birth place of economic organizations capable of manipulating supply and demand. But since in the marxist view trade and credit do not produce value then cities that engaged in those activities could not have possibly given rise to capitalism. As Deleuze and Guattari put it:

There is therefore an adventure specific to the towns in the zones where the most intense decoding occurs, for example, the ancient Aegean world or the Western world of the Middle Ages and the Renaissance. Could it not be said that capitalism is the fruit of the towns, and arises when an urban recoding tends to replace State overcoding? This, however, was not the case. The towns did not create capitalism. *The banking and commercial towns being unproductive* and indifferent to the backcountry, did not perform a recoding without also inhibiting the general conjunction of decoded flows. ... [Therefore it] was through the State-form not the town-form that capitalism triumphed. [21]

It is true that certain governmental organizations (not "the State") were instrumental in the creation of industrial organization as we know it today, since the industrial discipline and routinization of labor necessary for economies of scale are of military origin, born in armories and arsenals in France and the United States. [22] But that is an entirely different question, having more to do with an overcoding of labor than with the decoding effect of prices set by supply and demand. This is why locating assemblages at the right level of scale, a population of organizations that includes military ones, in this case, is so important. It is also necessary to stick to an ontology without reified generalities. Unfortunately, much of the academic left today has become prey to the double danger of abandoning materialism and of politically targeting reified generalities (Power, Resistance, Capital, Labor). A new left may yet emerge from these ashes but only if it recovers its footing in a mind-independent reality and if it focuses its efforts at the right social scale. This is where materialist philosophers can one day make a difference.

REFERENCES:

1. Michel Foucault. Discipline and Punish. (New York: Vintage Books, 1979), p. 195-199.

2. Gilles Deleuze and Felix Guattari. A Thousand Plateaus. (Minneapolis: University of Minnesota Press, 1987), p. 40-41. (Italics in the original)

3. Gilles Deleuze and Felix Guattari. Ibid. Pages 59-60. (Italics in the original)

4. Ibid. p. 62.

5. Gilles Deleuze and Claire Parnet. Dialogues II. (New York: Columbia University Press, 2002), p. 124 and 135.

6. See the previous essay in this book.

7. Gilles Deleuze. Foucault. (Minneapolis: University of Minnesota Press, 1988), p. 47.

8. Max Weber. The Theory of Social and Economic Organization. (New York: Free Press of Glencoe, 1964), p. 331.

9. Ibid. p. 328-336.

10. Ibid. p. 348.

11. Ibid. p. 359.

12. Michel Foucault. Discipline and Punish. Op. Cit. p. 169.

13. Charles Tilly. Stories, Identities, and Political Change. (Lanham: Rowman and Littlefield, 2002), p. 28-29.

14. Ibid. p. 90.

15. Ibid. p. 12.

16. Gilles Deleuze and Felix Guattari. A Thousand Plateaus. Op. Cit. p. 454-455.

17. Fernand Braudel. The Perspective of the World. (Harper and Row, New York, 1986), p. 630.

18. Ibid. p. 632.

19. Paul M. Hohenberg and Lynn H. Lees. The Making of Urban Europe, 1000-1950. (Cambridge: Harvard University Press, 1985), p. 281-282.

20. Gilles Deleuze and Felix Guattari. A Thousand Plateaus. Op. Cit. p. 463.

21. Ibid. p. 434. (My italics.)

In the sentence following this quote Deleuze and Guattari go on to quote Braudel arguing that territorial states eventually won the race against city states. This is, of course, true but not because the former allowed an axiomatic to flourish but because they had a larger population and hence not only a much bigger reservoir of recruits for their armies, but also a more extensive tax base with which to fund the increasingly expensive arms races. The only conclusion we can draw from this is either that Deleuze and Guattari did not read all three volumes of Braudel's history, or that if they did, they did not understand the implications.

22. David A. Hounshell. From the American System to Mass Production. 1800-1932. (Baltimore: John Hopkins University Press, 1984). Chapter 1.

Assemblage Theory and Linguistic Evolution.

Approaching the study of language from the point of view of assemblage theory is a difficult task because linguistic entities operate at several levels of scale at once.   First, words and sentences are component parts of many social assemblages, such as interpersonal networks and institutional organizations, interacting not only with the material components of those assemblages but also with non-linguistic expressive components. Second, some linguistic entities (religious discourses, written constitutions) have the capacity to code all the components of a given assemblage. While in the first case linguistic entities are variables of a social assemblage in the second case they become a parameter of it. Finally, language itself may be studied as an assemblage, exhibiting the characteristic part-to-whole relation: sounds or letters interact to form wholes, words, with irreducible semantic properties of their own; words interact to form larger wholes, sentences, with their own semantic and syntactic properties, and so on.

The expressive components of language as an assemblage include not only the meanings of words and sentences, but also other non-semantic sources of expressivity: tone, stress, rhythm, rhyme. The material components are either acoustic matter – pulses of air produced in the larynx and shaped by tongue and palate, teeth and lips – or physical inscriptions: carvings on stone, ink on paper, or the ones and zeroes that code language into electricity flows in the Internet. When studying language itself as an assemblage it is important to emphasize that there is no such thing as "language in general". That is, we must replace the reified generality "Language" with a population of individual singularities: a plurality of individual dialects coexisting and interacting with individual standard languages.

Although for analytical purposes it is convenient to distinguish these three different levels – language as a variable of social assemblages, language as a parameter, and language itself as an assemblage – it is clear that in any concrete case all three levels will operate simultaneously and influence one another. In this essay I will describe the three levels separately, starting with the role of language as a variable, or as an expressive component of assemblages like communities and organizations. To locate the analysis at the correct level it is important to distinguish *properties from capacities*. While the phonological, semantic, and syntactic properties of a language remain roughly constant at different levels, its capacities do not. The distinction between properties and capacities is roughly that between *what an entity is and what it can do*. In a famous text, appropriately named "How to Do Things With Words", the philosopher John Austin introduced the idea that sentences (or more exactly, statements) have the capacity to perform acts that create social commitments by their very utterance: promises, bets, apologies, threats, warnings, commands, death sentences, war declarations, and a large variety of other "speech acts". [1]

Speech acts are often performed by statements that have certain semantic and syntactical properties: performative statements like "I promise that X. " or "I command you to do Y." But these properties are not sufficient to endow them with their (illocutionary) capacities. Take for example a judge's utterance of a death sentence: "I declare you guilty and condemn you to die." This statement has a capacity to affect only if the person who utters it is someone with legitimate authority that is part of the right organization: a judicial, and not a legislative or executive, governmental organization. Moreover, the statement must be addressed to someone with the capacity to be affected. Animals other than humans, for example, cannot be declared guilty in a court of law, and in many judicial organizations a mentally ill person cannot be condemned to die. Commands too, presuppose the existence of an authority hierarchy, like the chain of command of a military organization. For the speech act to have binding capacities those who give orders as well as those who receive them must have the right positions in that hierarchy: one must be a subordinate and the other his or her legitimate superior. Other speech acts presuppose not the existence of organizational assemblages but of communities. In a tightly-knit community, for example, word of mouth about dishonored commitments travels fast and these violations can be punished by ridicule or ostracism. Hence, a community is the assemblage within which promises, bets, apologies, assertions, have a capacity to affect, and only its members have the capacity to be affected.

How exactly do speech acts affect a member of an organization or a community? By changing his or her social status. After a promise has been made, a bet agreed upon, an apology accepted, the status of a community's member is changed by that event in the sense that he or she is now regarded by other members as *having acquired a specific social obligation*: to keep the promise, to pay the bet, to behave in a forgiving way. And similarly for speech acts performed in organizations: a guilty verdict changes the status of someone from innocent to guilty with all the consequences this has; and once an order has been given in a military organization the status of the soldier who received the command is changed by it, having the obligation to carry out the command or be punished for insubordination. Deleuze and Guattari distinguish between the events in which the capacities of language are actualized, the *incorporeal transformations* they effect as expressive components, from the material components of an assemblage. As they write in relation to a judges' sentence:

> In effect, what takes place before hand (the crime of which someone is accused), and what happens afterward (the carrying out of the penalty), are actions-passions affecting bodies (the body of the property, the body of the victim, the body of the convict, the body of the prison); but the transformation of the accused into a convict is a pure instantaneous act or incorporeal attribute that is the expressed of the judge's sentence. Peace and war are states or interminglings of very different kinds of bodies, but the declaration of a general mobilization expresses an instantaneous and incorporeal transformation of bodies. [2]

Deleuze and Guattari take the concept of a speech act from Austin and add some elements of their own. Specifically, they invent the concept of an *order-word* to connect speech act theory to the theory of assemblages. On one hand, order-words refer to the capacities of statements to create commitments: "Order-words do not concern commands only, but every act that is linked to statements by a social obligation. Every statement displays this link, directly or indirectly. Questions, promises, are order-words." [3] On the other hand, they argue that the concept of an order-word is not related to the communicative functions of language, functions that imply the existence of a subject or person with intentions to communicate, but to the *impersonal transmission* of statements (and their illocutionary capacities). When a command is given in a military hierarchy, for example, the intentions of the commander issuing the order may be important but from then on it is the transmission of the order all along the chain of command that matters. And when the dishonoring of a commitment is witnessed in a community, the intentions of the person witnessing a broken promise or a false assertion are only the start of a process of transmission via word of mouth, an impersonal process that gives the community as a whole its capacity to store the reputations of its members. As they write:

> If language always seems to presuppose itself, if we cannot assign it a non-linguistic point of departure, it is because language does not operate between something seen (or felt) and something said, but always goes from saying to saying. We believe that narrative consists not in communicating what one has seen but in transmitting what one has heard, what someone else said to you. Hearsay. ... Language is not content to go from a first party to a second party, from one who has seen to one who has not, but necessarily goes from a second to a third party, neither of whom has seen. It is in this sense that language is the transmission of the word as order-word, not the communication of a sign as information. [4]

A community (or organization) possessing as one of its expressive components transmissible order-words is referred to as a "collective assemblage of enunciation". [5] Order-words are one of the variables characterizing these assemblages, variables capable of taking as many values as there are possible speech acts. But how are we to conceive of other, larger linguistic entities that seem to code every component of these assemblages? In the case of the judge's sentence, for example, the court (as part of a larger assemblage of judicial organizations) is not only characterized by discrete events in which statements exercise their capacities to affect but also by a larger body of law that codifies past legal knowledge and precedent. It is by referring to that legal code that a particular sentence acquires its legitimacy. Similarly, in a military organization (and most civilian organizations) there will be a written constitution codifying the functions, rights, and obligations granted by the government to that organization. This constitution forms the background against which particular commands can be given with legitimate authority. Although these written codes also operate, on a day-to-day basis, through the transmission of order-words, they are not so much variables of these assemblages as much as parameters quantifying their overall environment. And similarly for communities: a religious discourse, codifying places into sacred and profane, food into permissible and taboo, days of the year into ordinary and special, affects every expressive and material component of the social assemblage.

The distinction between variables and parameters comes from mathematical models of physical processes. Temperature can appear in a model as a variable, the internal temperature of an animal's body, for example, as well as a parameter quantifying the degree of temperature of the animal's surroundings. We can borrow this distinction to capture the role played by linguistic entities when they affect all components of an assemblage, that is, when they are part of the environment of an assemblage. We can go even further and *parametrize* the concept of assemblage, building into it "control knobs" with variable settings: the values of one knob can quantify the degree to which the components of the assemblage are uniform or the degree to which its defining borders are sharp (the territorialization parameter); the other knob can quantify the extent to which linguistic categories belonging to a legal code or religious discourse have been systematically assigned to these components (the coding parameter.) The advantage of a parametrization is not only that it permit us to capture the inherent variability of the identity of all historically constituted entities, but also that it allows us to easily deploy the concept in our minds in large numbers, forming a conceptual population in which the variation possesses a certain statistical distribution. [6]

Being able to think about entire assemblage populations, and about the statistical form in which their variation is distributed, is crucial to the application of assemblage theory to linguistic evolution. Although the above discussion emphasizes the transmissibility of order-words among the members of organizations or communities, it is clear that order-words are also transmitted across generations, from parents (or teachers) to children. In this respect words and sentences are similar to genes, that is, they are *replicators*. Biologists have known for a while that genes do not have a monopoly on evolution. Learned patterns of behavior, that is, behavioral patterns that are not genetically hard-wired, can be transmitted from one generation of animals to another by imitation. The best studied example of this evolutionary process is bird songs, like the complex songs of nightingales or blackbirds. These replicating behavioral patterns are referred to as "memes". [7] In human populations memes are exemplified by fashion (dress patterns, dancing patterns) but a more important type of replicating entity, one that replicates not by imitation but by enforced social obligation, is the sounds, words, and grammatical patterns that make up a language: although babies may at first aim at imitating the sounds coming out of their parents mouths, they soon learn that speaking their mother tongue is not optional but obligatory, and that there is a norm (the dialect spoken in their community) to which they must conform.

In all cases of evolution variation is indispensable. If genes replicated exactly, if there were no copying errors or mutations, organic entities would not have the capacity to evolve because the selection pressures that promote the replication of some variants at the expense of others, leading to the slow accumulation of adaptive traits, would have no raw materials on which to operate. And this is true too of any other kind of replicating entity. In the case of dialects the amount of variation, and the extent to which it is centripetal or centrifugal (territorializing or deterritorializing), is in many cases determined by the intensity of enforcement. Enforcement can be performed by an institutional organization, like the Academies of Language that, starting in the late sixteenth century in Europe, were chartered to create standard versions of the dialects of dominant regional capitals, like Florence or Paris, and to enforce them through the publication of official dictionaries, formal grammars, and books of correct pronunciation. But enforcement can also be performed by tightly-knit communities. When sociolinguists try to account for the survival of so many dialects of English or French into the twentieth century – despite the homogenizing influence of compulsory primary education in the standard, of radio and television broadcasting in the standard, of the spread of tourists from the national capitals – they use two explanatory factors. First, language is used by community members not only to communicate but also as a badge of identity. [8] A young member of a small town that goes to a big city to study and returns with a different accent will not be treated with respect, as bringing a more sophisticated variant to enlighten the locals, but as an outsider. Second, the capacity of tightly-knit communities to store reputations and detect violations of local norms, coupled to the use of ridicule and ostracism as forms of punishment, gives them the capacity to enforce and preserve their local linguistic identity. [9]

Let's illustrate this with a few examples. Before the collapse of the Roman empire, all the regions the Romans had conquered, and on which they had imposed Latin as the official language, were relatively linguistically homogenous. The area under Roman control was so vast that there must have been local variations, but the very presence of Roman troops and officials (as well as of governmental organizations) kept the variation centripetal: to be able to address a government official, for example, the locals had to use the Latin of Rome, and this gave them a standard to use as a norm. But the moment the empire ceased to exist, and with it the organizations that enforced its rule, the variation turned centripetal, as local communities replaced imperial organizations as enforcement mechanisms. Vulgar Latin, the Latin spoken by the conquered masses, began to change and gave birth to a large variety of Romance languages. [10] At first, this explosive divergence went unnoticed: people in the Latinized areas of the former empire thought they were all speaking Latin, and there were hardly any names for the emerging variants. Awareness of the existence of new versions of Vulgar Latin involved an organizational intervention: the reforms that the court of Charlemagne introduced in the early Ninth century. A professional grammarian, named Alcuin, was hired to report on the state of language in the kingdom, and he informed the king that something new existed outside the walls of the castle, a new language he called "Rustica Romana". [11]

Many Romance dialects coexisted in the Middle Ages in what is called a "dialect continuum". [12] The dialect of medieval

Paris, for example, was connected to the dialect of Florence by a continuum of French, Franco-Provencal, and Gallo-Italian dialects. Although relatively sharp transitions (isoglosses) did exist in this continuum, compared to the form of Latin spoken in governmental and ecclesiastical organizations, the divergent set of Franco-Romance, Hispano-Romance, and Italo-Romance dialects was a highly deterritorialized entity. On the other hand, prestigious Latin, taken from classical books and given a spoken form during the Carolingian reforms, was highly territorialized and coded: its internal homogeneity could be preserved by reference to Roman texts; its borders policed by aristocratic and religious communities; and its uses (reading the Bible aloud in mass, writing laws and edicts) codified. To apply assemblage theory to these entities Deleuze and Guattari introduced the concepts of *major and minor languages*. As they write:

> Must a distinction be made between two kinds of languages, "high" and "low", major and minor? The first would be defined precisely by the power of constants, the second by the power of variation. We do not simply want to make an opposition between the unity of a major language and the multiplicity of dialects. Rather, each dialect has a zone of transition and variation... [It] is rare to find clear boundaries on dialect maps; instead there are transitional and limitrophe zones, zones of indiscernibility. ... The very notion of dialect is quite questionable. Moreover, it is relative because one needs to know in relation to what major language it exercises its function: for example, the Québecois language must be evaluated in relation to standard French but also in relation to major English, from which it borrows all kinds of phonetic and syntactical elements, in order to set them in variation. ... In short the notion of dialect does not elucidate that of a minor language, but the other way around; it is the minor language that defines dialects through its own possibilities of variation. [13]

The relativity of the notions of major and minor language can be captured by a parametrized concept in which the values of the parameters can vary historically. Thus, while all the vernacular forms of Romance continued to be minor relative to major classical Latin in the first half of the second millennium, some dialects became major relative to others. The commercial revolution that took place in Europe from the eleventh to the fourteenth centuries, and the diversification of governmental functions in the proliferating city states, multiplied the uses of writing: licenses, certificates, petitions and denunciations, wills and post-mortem inventories, commercial and financial contracts began to be written with increased frequency. [14] Because this rising demand was not matched by an equivalent supply of classical Latin scribes the governments of several regional capitals commissioned the creation of writing systems for their own dialects. [15] The appearance of a written form had several consequences: it reduced the intensity of variation in those dialects, decelerating their evolution relative to those without writing; it increased awareness of their unique identity among their users; and it augmented their level of prestige relative to the main major standard. In short, writing had territorializing effects that made some members of the continuum more discrete and constant, transforming them into major languages relative to the remaining minor ones.

The evolutionary process that gave birth to the ancestors of today's English dialects shows a similar pattern. The raw materials for this process were brought to England, starting in the Fifth century, by several migratory waves from Germanic speaking Jutes, Angles, and Saxons. These created a dialect continuum in the island similar to the one that was forming in the continent. One of these dialects, what we today call "West Saxon", was given a writing system and its prestige was greatly increased with the appearance of the Beowulf, a literary masterpiece written in that dialect. Thus, West Saxon became a major language in relation to the many other minor variants. But then, French-speaking Normans staged a brutal invasion in which the English governing elite was physically exterminated, and with it any organizational means to enforce the rising standard. A defeat at the Battle of Hastings in 1066 sealed not only the fate of that elite but also became a transformative event in the history of their language. In the words of historian John Nist:

As a result of the Norman conquest, the Old English nobility practically ceased to exist. Within ten years after the Battle of Hastings the twelve earls of England were all Norman. Norman clergy ... took overs the highest offices of the Church: archbishop, bishop, and abbot. Since the prestige of a language is determined by the authority and influence of those who speak it, the French of the Norman masters became the tongue of status in England for almost two hundred years. The Norman King of England and his French nobility remained utterly indifferent to the English language until about 1200. During that time the royal court patronized French literature, not English. When the court and its aristocratic supporters did finally pay attention to the native language of the land they dominated, that language was no longer the basically Teutonic and highly inflected Old English but the hybrid-becoming, Romance-importing, and inflection-dropping Middle English. [16]

Here we have an example of a major language (West Saxon) being replaced by another (Norman French), removing all territorializing and coding pressures on a population of minor languages and thereby accelerating their evolution. Of all the different changes that the population underwent perhaps the most important was the loss of inflections. An inflection is a syllable that, as part of a word, carries with it grammatical information. Highly inflected languages, whether Germanic or Romance, use these syllables to express gender and number in nouns, and person and tense in verbs. The English peasants that lived during the Norman occupation had inherited the habit of placing stress on the very first syllable. This habit can still be detected in contemporary English words from different origin. Thus, words of Germanic origin consistently stress the root syllable (as in   "love" "lover", "loveliness") while those borrowed from Romance do not: "family", "familiar", "familiarity". Without the enforcement of any norms beyond those of local communities, and given this habitual stress, the last syllables became literally eroded away. And without inflections English dialects were forced to use a relatively fixed word order to express grammatical function, a major structural transformation. As Nist puts it: "Unhindered by rules of proscription and prescription, the English peasants demanded stress in the root syllable and remodeled the language with tongue and palate." [17]

What kind of linguistic models do we need to accommodate these historical metamorphoses? Or to put this differently, what does language have to be, what kind of properties must it have, in order for it to possess these transformative capacities? Some linguistic models cannot account for evolutionary processes, or rather, subordinate the evolution of language to that of the human species. Chomsky's model, for example, postulates the existence of a universal grammar, a constant core common to all languages, evolved genetically and residing in our brains. This innate universal grammar includes linguistic categories (sentence, noun, verb) as well as re-writing rules that can transform one string of words belonging to those categories into another string. [18] But in order to model an evolutionary process in which the replicators are linguistic, not genetic, we must get language out of our heads, and more importantly, the relations between words in a sentence must be conceived as *relations of exteriority not relations of interiority*. This distinction plays a crucial role in assemblage theory: a relation of interiority is one in which the terms constitute each other by the very fact that they are related. In other words, the terms of the relation do not have an autonomous existence independently of their relation. Hegelian totalities have that character, as do grammatical relations in Chomskian linguistics: nouns and verbs are constituted by their very relation in a sentence, as part of the totality of a universal grammar. Relations of exteriority, on the other hand, link terms that exist independently of their being related. The interactions between components of any assemblage, interactions in which they exercise their capacities to affect and be affected, are of this type.

A model of language that meets this requirement has been created by Zellig Harris. In this model, words carry with them in addition to their semantic information, that is, their meaning, non-linguistic information about their *frequency of co-occurrence* with other words. In this second sense, the term "information" refers to physical patterns, like the patterns of ones and zeroes processed by computers, so it is a measure of order, or of the degree to which a pattern departs from randomness. The non-linguistic information carried by words reflects the fact that some words tend to occur next to other words more frequently as a matter of actual usage. Thus, after a speaker has uttered a definite article like "the" the listener can reasonably expect that the next word will be a noun or a nominalized phrase. Harris calls these relations "likelihood constraints". At any given point of time, these constraints may simply reflect the word combinations that happen to be used in a community (and hence be optional) but these customary patterns may eventually become standardized or conventionalized and turn into obligatory constraints: a user who starts a sentence with "the" may now be required to supply a noun or nominalized phrase as the next word. [19] Saying that a likelihood constraint becomes obligatory is not, of course, to imply that the words themselves enforce the social obligation, but rather that the words are used in a collective assemblage of enunciation, like a tightly-knit community, possessing enforcement capabilities.

Modeling words as carrying statistical information about their frequency of co-occurrence provides a non-genetic evolutionary mechanism for the emergence of word categories. When a word is used very frequently after another word, listeners come to expect with very high confidence that the second word will occur after hearing the first word. This implies that uttering the second word may become *redundant* since the speaker can count on the listener to provide it. Harris calls this "reduction constraints". These constraints can reduce whole words into suffixes or prefixes attached to another word, or even zeroed altogether. Moreover, entire sentences may be compacted, by eliminating all redundant words, while preserving the full semantic content of the uncompacted sentence, because this meaning is reconstructed by the listener. As Harris shows, successive applications of reduction constraints, particularly when they have become obligatory, can give rise to new classes of words, like adjectives, adverbs, conjunctions, prepositions. [20] In his view the earliest forms of language evolved starting from monolithic symbolic artifacts, that is, symbolic artifacts that did not have the capacity to combine into larger linguistic entities. At first, these monolithic artifacts had the same probability of occurring, that is, their relative frequency was random. But then the original symmetry was progressively broken by *successive departures from equiprobability*, changing the co-occurrence patterns and slowly endowing the artifacts with combinatorial capabilities.

Once classes of words established themselves a final constraint could emerge: the operator-argument constraint, establishing obligatory relations between those classes. This final constraint is needed to model the action of adjectives on nouns or of adverbs on verbs, that is, to model the capacity that some words have to modify other words. The operator-argument constraint is also informational in the non-linguistic sense: the more unfamiliar the argument supplied for a given operator, the more informative it is; and vice versa, an argument that becomes too familiar becomes redundant and it can become the target for a reduction constraint. Deleuze and Guattari stressed the importance of redundancy (and of frequency, as a specific form of redundancy) in their theory of order-words, but did not provide us with a model of language in which redundancy played a morphogenetic role. [21] The work of Zellig Harris, on the other hand, provides us with the means to plug this hole in the assemblage approach to language. His model not only treats words as material entities entering into relations of exteriority with one another, but it is an explicitly evolutionary model: language evolves in a non-genetic way through the obligatory social transmission of combinatorial constraints, as words and sentences compete for "informational niches". [22]

Another shortcoming in Deleuze and Guattari's analysis is the use of the term "collective assemblage of enunciation" as a general term. In this essay I have tried to correct this tendency by always referring to concrete social assemblages, communities or organizations, but these are not the only cases. A social justice movement, for example, being a coalition of many communities, is also a collective assemblage of enunciation in which order-words take the form of slogans and other expressions of unity and purpose. To be able to extract concrete rights from a government, or to consolidate their gains, these movements may also have to add to the assemblage one or more organizations. Similarly, although the academies of language that created standard languages in many Romance speaking countries had a powerful territorializing linguistic effect, standard languages did not spread through entire countries without the help of much larger assemblages comprising many organizations: the network of schools that, starting in the nineteenth century, implemented compulsory primary education in the standard. Hence, to explain a particular historical episode we need to use concrete cases of collective assemblage of enunciation, not the general term. It is only by sticking to concrete assemblages – individual communities, individual organizations, and so on – that we can avoid introducing reified generalities into a historical analysis.

And it is only by eliminating reified generalities that an approach to history can become truly materialist.

REFERENCES:

1. John L. Austin. How to Do Things with Words. (Cambridge Mass: Harvard University Press, 1975), p. 26.

2. Gilles Deleuze and Felix Guattari. A Thousand Plateaus. (Minneapolis: University of Minnesota Press, 1987), p. 80-81.

3. Ibid. p. 79.

4. Ibid. p. 77.

5. Ibid. p. 88.

To this it must be added that communities and organizations are also an intermingling of heterogenous material bodies (humans, machines, tools, food and water, buildings and furniture, and so on). Hence, a community or organization is, on the one hand, "a *machinic assemblage* of bodies, an intermingling of bodies reacting to one another; on the other hand, it is a *collective assemblage of enunciation*, of acts and statements, of incorporeal transformations attributed to bodies.". (Italics in the original.)

6. On the idea of parametrizing the concept of assemblage see the first essay of this book.

7. Richard Dawkins. The Selfish Gene. (New York: Oxford University Press, 1990), p. 19-20

8. William Labov. The Social Setting of Linguistic Change. In Sociolinguistic Patterns. (Philadelphia: University of Pennsylvania Press, 1972), p. 271.

9. Lesley Milroy. Language and Social Networks. (Oxford: Basil Blackwell, 1980), p. 47-50

10. Alberto Varvaro. Latin and Romance: Fragmentation or Restructuring?. In Latin and the Romance Languages in the Early Middle Ages. Edited by Roger Wright. (London: Routledge, 1991). p. 47-48.

11. Roger Wright. The Conceptual Distinction between Latin and Romance: Invention or Evolution?. In Ibid. p. 109.

12. M.L. Samuels. Linguistic Evolution. (Cambridge University Press, London 1972), p. 90.

13. Gilles Deleuze and Felix Guattari. A Thousand Plateaus. Op. Cit. p. 101-102.

14. Peter Burke. The Uses of Literacy in Early Modern Italy. In Social History of Language. Edited by Peter Burke and Roy Porter. (Cambridge: Cambridge University Press, 1987), p.22-23.

15. Roger Wright. The Conceptual Distinction between Latin and Romance. Op. Cit. p. 104-105.

16. John Nist. A Structural History of English. (New York: St. Martin's Press, 1976), p. 106-107.

17. Ibid. p. 148.

18. Noam Chomsky. Aspects of the Theory of Syntax. (Cambridge: MIT Press, 1965), p. 66-73.

19. Zellig Harris. A Theory of Language and Information: A Mathematical Approach. (Oxford: Clarendon Press, 1981), p. 367.

20. Ibid. p. 339.

21. Gilles Deleuze and Felix Guattari. A Thousand Plateaus. Op. Cit. p. 79.

22. Zellig Harris. A Theory of Language and Information. Op. Cit. p. 324-326.

Metallic Assemblages.

The political, economic, and social regime of the peoples of the steppe are less well known than their innovations in war, in the areas of offensive and defensive weapons, composition or strategy, and technological elements (the saddle, stirrup, horseshoe, harness etc.) History contests each innovation but cannot succeed in effacing the nomad traces. What the nomads invented was the man-animal-weapon, man-horse-bow assemblage. Through this assemblage of speed, the ages of metal are marked by innovation. The socketed bronze battle-ax of the Hyksos and the iron sword of the Hittites have been compared to miniature atomic bombs. ... It is commonly agreed that the nomads lost their role as innovators with the advent of firearms, in particular the cannon. ... But it was not because they did not know how to use them. Not only did armies like the Turkish army, whose nomadic traditions remained strong, develop extensive firepower, a new space, but additionally, and even more characteristically, mobile artillery was thoroughly integrated into mobile formations of wagons, pirate ships etc. If the cannon marks a limit for the nomads, it is on the contrary because it implies an economic investment that only a State apparatus can make (even commercial cities do not suffice).

Gilles Deleuze and Felix Guattari. A Thousand Plateaus. [1]

The whole composed of a human being, a fast riding horse, and a missile-throwing weapon like the bow, is probably the best known example of an assemblage of heterogenous elements, cutting as it does across entirely different realms of reality: the personal, the biological, and the technological. This emergent whole is itself composable into larger assemblages: a nomad army, an assemblage of mobile cavalry formations in which the components can fight alone or coalesce into teams, variably adjusting to the conditions on the battlefield. By contrast, the kind of military assemblage that sedentary peoples created, such as the phalanx, was an inflexible block of infantry soldiers that could exercise no initiative in the battlefield and was therefore hard to control once the order to attack had been given. This distinction between nomad and sedentary military assemblages, one more flexible or *deterritorialized*, the other more rigid or *territorialized*, should not be taken to imply the existence of two general categories of armies. Rather, the degree of territorialization or deterritorialization should be treated as a variable parameter that can change historically.

The sedentary armies of Europe, for example, underwent a gradual "nomadization" in the last four hundred years: their phalanxes were first flattened in the sixteenth century, from the original eight-men deep formation to three-men deep; they were then given more flexibility during the Napoleonic wars; and they were finally broken down during World War II into relatively autonomous platoons capable of making tactical decisions on their own. This process of deterritorialization was caused by the steady pressure of more powerful and mechanized fire weapons, like the rifle and the machine gun, that made fighting in tight formations increasingly costly, as well as by the availability of portable radio allowing the articulation of many platoons through a wireless chain of command. [2]

An army, sedentary or nomadic, should be viewed as an *assemblage of assemblages*, that is, as an entity produced by the recursive application of the part-to-whole relation: a nomad army is composed of many interacting cavalry teams, themselves composed of human-horse-bow assemblages, in turn made out of human, animal, and technical components. Similarly, and simplifying a bit, a modern army is composed of many platoons, composed of many human-rifle-radio assemblages, the human and technical components of which are themselves assemblages. At any level of such a nested set of assemblages causality operates in two directions at once: the bottom-up effect of the parts on the whole, and the top-down effect of the whole on its parts. On one hand, the properties and capacities of a whole emerge from the causal interactions between its parts: many human-horse-bow assemblages, trained intensively to work together, form a whole with the emergent capacity to take advantage of spatial features of the battlefield, for ambush and surprise, and to exploit temporal features of the battle, such as the fleeting tactical opportunity presented by a temporary break in an enemy's formation. Because of this bottom-up causality the emergent properties and capacities of a whole are *immanent*, that is, they are irreducible to its parts but do not transcend them, in the sense that if the parts stop interacting the whole itself ceases to exist, or becomes a mere aggregation of elements. On the other hand, once a whole emerges it can exercise its capacities not only to interact with other wholes, as when two enemy armies face each other in battle, but to affect its own parts *constraining them and enabling them*. Belonging to a team of warriors makes its members subject to mutual policing: any loss of nerve or display of weakness by one member will be noticed by the rest of the team and affect his or her reputation. But the team also creates resources for its members, as they compensate for each other's weaknesses and amplify each other's strengths.

The existence of bottom-up and top-down forms of causality implies that the evolution of the components of an assemblage, at any given level of scale, will be partly autonomous and partly influenced by the environment created by the larger assemblage itself. Whether a particular technical object is used as a weapon or as a tool, for example, is in some cases determined by top-down causality. Let's take the example of a community of hunter-gatherers, that is, the communal assemblage of which human beings (and its pre-modern ancestors) were component parts for hundreds of thousands of years. In this assemblage there was not only a division of labor but also communal monitoring of behavior: a community as a whole could enforce a more or less egalitarian distribution of the spoils of the hunt. This means that even though for most of this time humans did not possess language they were already fully social beings, not to mention highly skilled producers of stone artifacts. We can imagine that in the daily life of a hunter-gatherer community these stone artifacts were tools, used to hunt as well as to butcher and skin the carcass, but that when one community faced another in violent conflict, the same object became a weapon. That is, the object's properties remained the same but it was used in a very different way: not directed with controlled movements towards a carcass, but projected towards, or even thrown at, an enemy. In other words, the communal assemblage selected some capacities of the stone artifacts when in "work mode", and other capacities when in "war mode". As Deleuze and Guattari write:

... the principle behind all technology is to demonstrate that a technical element remains abstract, entirely undetermined, as long as one does not relate it to an assemblage it presupposes. It is the machine that is primary in relation to the technical element: not the technical machine, itself a collection of elements, but the social or collective machine, the machinic assemblage that determines what is a technical element at a given moment, what is its usage, its extension, its comprehension, etc. [3]

Expressing the reality of top-down causality this way is problematic because it seems to deny any autonomy to the components of an assemblage. As the above quote continues, for example, Deleuze and Guattari assert that "technical objects have no distinctive intrinsic characteristics". [4] If by that one means that there is not a set of necessary and sufficient characteristics that unambiguously characterizes weapons and tools in general then I agree with that assertion. Entities like "the Weapon" and "the Tool" are only reified generalities and as such have no place in assemblage theory. But expressing this by saying that a technical object outside a larger assemblage remains entirely undetermined risks transforming the concept of assemblage into something like a Hegelian totality, in which the very identity of the parts is constituted by their relations in the whole. To avoid this danger it is important to distinguish two different ways in which technical objects may be characterized: by their properties and by their capacities.

Let's use a knife as an example. Its properties include its length, weight, and sharpness. These properties characterize the more or less enduring states of the knife and are therefore always actual: at any one point in time a knife is either sharp or blunt. A sharp knife, on the other hand, also has capacities, like its capacity to cut. Unlike sharpness, the capacity to cut need not be actual, if the knife is not presently cutting something, and may never become actual if the knife is never used. And when a capacity does become actual it is never as an enduring state but as a more or less instantaneous event. Moreover, this event is always double, *to cut-to be cut*, because a capacity to affect must always be coupled to a capacity to be affected: a particular knife may be able to cut through bread, cheese, paper, or even wood, but not through a solid block of titanium. This implies that while properties are finite and may be put into a closed list, capacities to affect may not be fully enumerated because they depend on a potentially infinite number of capacities to be affected. Thus, a knife may not only have a capacity to cut but also a capacity to kill, if it happens to interact with a large enough organism with differentiated organs, that is, with an entity having the capacity to be killed.

The assertion that a particular technical object is a tool or a weapon depending on the larger assemblage of which it is a part could then be interpreted as meaning that a knife as used in a "kitchen assemblage" is a tool, the assemblage selecting from all its capacities only the ability to cut, while the same knife in an "army assemblage" becomes a weapon, the assemblage selecting its ability to kill. Yet, this would not imply that the properties of the knife are determined by the larger assemblage. Those properties, on the contrary, emerge from the interactions between a knife's own components. The sharpness of its blade, for example, is a geometric property of the cross-section of the blade (its triangular or pointy form) a property that emerges from a particular arrangement of its component crystals. And similarly for larger assemblages: a sword, a human being, and a horse, must exercise certain causal capacities as they interact with each other for the man-horse-weapon assemblage to have emergent properties: the human must ride the horse to acquire its momentum (its larger mass multiplied by its faster speed), and must hold a weapon firmly in hand to transmit that momentum to it. But once the assemblage emerges, it constrains its components discarding some capacities – the capacity of the rider to be compassionate; the capacity of the horse to pull carriages; the capacity of the sword to be used as a tool – and selecting others: fighting skills, rapid movement, projected motion.

Whereas the stone artifacts of hunter-gatherers may indeed be viewed as relatively undetermined, the same object having two different uses within a community and between conflicting communities, once tools and weapons differentiated in form and function they acquired a certain autonomy. No doubt the progressive differentiation of technical objects was guided by the differentiated social assemblages (farms, armies,

workshops, temples) of which they were parts. But they
nevertheless retained their own properties, a fact that explains,
for example, that they could be detached from one assemblage
and plugged into another, as when fire weapons were transfered
from sedentary armies to nomad ones. And technical objects can
be said to have their own history, in the sense that the pace of
technological development may pick up speed, accelerating
relative to that of institutional development and forcing the latter
to catch up. That seems to be the case with rifles and machine
guns, the lethal capacities of which demanded a break with tight
formations, a demand that went unsatisfied for more than a
century as armies lagged behind technology, unable to reform
themselves and adapt to the new conditions on the battlefield.

But while distinguishing properties from capacities may
help us to correctly interpret the quote above, there is another
problem with it that is not so easy to solve: the fact that Deleuze
and Guattari seem to be using two incompatible definitions of
the term "assemblage". In his own texts, Deleuze uses the term
to refer not only to social assemblages, like the man-horse-
weapon assemblage, but to biological ones (the wasp-orchid
symbiotic assemblage) and even non-organic ones, like the
assemblage formed by copper and tin when they interact to form
an alloy, bronze, with its own emergent properties and
capacities. [5] But in his joint work with Guattari the term refers
only to social assemblages. Thus, an army is both a "machinic
assemblage" or an intermingling of material bodies (human,
animal, technical bodies) as well as a "collective assemblage of
enunciation", that is, a whole in which statements have the
capacity to create social obligations, like the commands that
flow downwards in an army hierarchy, or the reports that flow
upwards. [6] In this second sense, the term "assemblage" is not
only restricted to social wholes, since only in them can
statements be speech acts, but it applies to only one level of
scale: a different term is used to refer to the components of an
assemblage, "bodies", and another term to refer to the larger
social wholes that assemblages form, "the social field" or "the
socius". Thus, whereas in the first definition the components of a
man-horse-weapon assemblage can themselves be viewed as
assemblages, in the second one they cannot.

In my own work I have solved this ambiguity by always using the first definition, while trying to capture the content of the second one through the distinction between the material and the expressive components of an assemblage, the latter including not only statements and the speech acts of which they are capable, but also components capable of non-linguistic forms of expression. Furthermore, while Deleuze includes in the first definition the requirement that the components be heterogenous (using a different term, "stratum", to refer to wholes with homogenous components) I have modified the concept by parametrizing it, that is, by building into it "control knobs" with variable settings quantifying the degree of homogeneity or heterogeneity of the components, or the degree to which the assemblage's identity is rigidly or flexibly determined. Thus, as the opening paragraph of this essay indicates, I do not treat nomad and sedentary armies as two different categories but as two assemblages in which this parameter has different settings, quantifying two different degrees of territorialization. As the parameter changed under the pressure of faster and more accurate firearms, the phalanx was "nomadized", that is, it was deterritorialized.

It may be objected that parametrizing the concept of assemblage cannot replace the dichotomy stratum/assemblage because, after all, variations in a parameter can cause only *quantitative* changes, whereas the use of two different categories indicates a commitment to the existence of *qualitative* differences. This is more an apparent than a real problem as can be seen by examining the use of parameters in mathematical models or experimental situations, in the realms of physics, chemistry, or biology. In these fields a parameter, when it quantifies an intensive property, is characterized by critical points of intensity that mark sudden changes in quality. Let's take the example of speed. In fluid materials changes in this intensive parameter causes qualitative changes in regime of flow: at slow speeds a fluid tends to move in a uniform manner called "laminar flow" but at a critical point in speed this regime is replaced by a qualitatively different one, a coherent circular motion called "convective flow", which is in turn replaced at a another critical point by a regime called "turbulent flow". These three regimes exemplify the emergence of qualitatively distinct *phases* that cannot be confused with general categories of flow. Turbulence, in particular, with its complex structure of vortices within vortices, is often used by Deleuze and Guattari as a model for a qualitatively distinct regime or zone of intensity. As they write:

> It is thus necessary to make a distinction between *speed* and *movement.* A movement may be very fast, but that does not give it speed; a speed may be very slow, or even immobile, yet it is still speed. Movement is extensive; speed is intensive. Movement designates the relative character of a body considered as "one", and which goes from point to point; *speed, on the contrary, constitutes the absolute character of a body whose irreducible parts (atoms) occupy or fill a smooth space in the manner of a vortex,* with the possibility of springing up at any point. (It is therefore not surprising that reference has been made to spiritual voyages effected without relative movement, but in intensity, in one place: these are also part of nomadism.) [8]

Phase transitions driven by a speed parameter also occur in biology. Take for example one of the components of the man-horse-weapon assemblage. As a quadrupedal animal, a horse's manner of moving, its *gait*, spontaneously undergoes qualitative changes at critical points of speed: at low-intensity values of this parameter the horse walks; at faster values it trots; and to reach even higher speeds the horse is forced to break into a gallop. [9] Speed is also important when analyzing larger assemblages, the difference in the quality of motion between the rapid charge of cavalry formations and the slow march of a phalanx being rather obvious. But the parameter we need to quantify degrees of territorialization and deterritorialization in military assemblages is a *complex function of speed and other variable properties.* For example, the parameter must also quantify the degree to which decision-making is centralized or decentralized in an army, since allowing warriors to display more initiative in the battlefield is what made nomad armies capable to take advantage of topographical variations in the battlefield and adjust to the temporal variability of the battle itself. In Napoleonic armies, for example, an increase in speed and mobility was coupled to a more flexible chain of command. The degree of intensity of centralization of command was also crucial for the emergence of the semi-autonomous platoon: thresholds of decision-making were lowered allowing troops on the ground to make tactical decisions, while their commanders set only overall strategic goals.

Replacing general terms (assemblage, stratum) with a single parametrized concept is justified in a materialist philosophy not only to purge reified categories from its ontology, but also because the different regimes or phases into which the space of possible parameter values is divided can be transformed into each other. This allows historical analyses to account for the fact that assemblages of soldiers can become assemblages of workers, as when an army besieging a city must become sedentary and dedicate most of its time to logistical labor. As Deleuze and Guattari put it: "But it is not impossible for weapons and tools, if they are taken up by new assemblages of metamorphosis, to enter into other relations of alliance. The man of war may at times form peasant or worker alliances, but it is more frequent for a worker, industrial or agricultural, to reinvent a war machine." [10]

In addition, replacing categories with regimes or phases can help eliminate certain paradoxical, or even problematic, aspects in Deleuze and Guattari's argument. In particular, they constantly express admiration for the war machine, particularly when they oppose it to the state apparatus, that is, to a highly territorialized (and coded) assemblage of government organizations. But this constant praise should not be taken to imply that they view any particular military assemblage – not even one composed of nomad warriors whose legendary cruelty is well known – as a model for a better social order, and certainly not to imply approval or commendation of war itself. Rather, as a name for a phase or a zone of intensity in the space of possible values for the control parameter, the term "war machine" refers to a special regime in the operation of any organizational assemblage, a regime in which the organization displays a capacity *to operate in continuous variation*.

Much as regimes of flow, like convection or turbulence, can exist in a wide variety of fluid substances, from air and water to molten rock, the regimes into which critical points break the parameter space for organizations possess a certain universality, that is, they can be actualized in different organizations. In fact, Deleuze and Guattari use the terms "war machine" and "nomadic" to refer to many other assemblages, including *bodies of knowledge* and the organizations in which they are produced. In this sense, for example, they speak of "nomad sciences" as fields of knowledge production that effectuate the war machine. [10] They contrast these fields to those of sedentary (or Royal) sciences that, they postulate, occupy a zone of intensity in parameter space neighboring that of the State apparatus, that is, a highly territorialized (and coded) regime.

Whereas sedentary fields of science search for the eternal and immutable laws of nature, and treat matter as an obedient and domesticated substrate that faithfully follows those laws, nomad sciences treat matter not as an inert receptacle for forms that come from the outside (as in the so-called "hylomorphic model") but as animated from within by its own *tendencies and capacities*. [11] Deleuze and Guattari use the term "affect" (or "affective quality") to refer to capacities to affect and be affected, and the term "singularity" to refer to tendencies, such as the tendency of iron to melt at 1535 degrees centigrade. The term "singular" is used here not as the opposite of "plural" but of "ordinary", in the sense that in a line of temperature values the points marking 1532, 1533, 1534 (and many other) degrees are ordinary, nothing special happening at those points, while 1535 degrees is remarkable or singular. Moreover, 1535 degrees centigrade is only a constant if we use a single parameter, temperature, but it becomes a variable once we add a second parameter, like pressure. Metallurgy is the example of a nomad science that Deleuze and Guattari discuss in most detail, partly because the blacksmith as a producer of weapons has had a long association with military assemblages, but also because metallurgy illustrates that special regime of deterritorialization that allows an organization (a workshop in this case) to feed on variation itself, as opposed to subsist on a diet of constants, routines, and homogenized material behavior. As they write:

It would be useless to say that metallurgy is a science because it discovers constant laws, for example, the melting point of a metal at all times and in all places. For metallurgy is inseparable from several lines of variation: variation between meteorites and indigenous metals; variation between ores and proportions of metal; variation between alloys, natural and artificial; variation between the qualities that make a given operation possible or that result from a given operation... All of these variables can be grouped under two overall rubrics: *singularities or spatiotemporal haecceities* of different orders, and the operations associated with them as processes of deformation or transformation; and *affective qualities or traits of expression* of different levels, corresponding to these singularities and operations (hardness, weight, color etc.). Let us return to the example of the saber, or rather, of crucible steel. It implies the actualization of a first singularity, namely, the melting of the iron at high temperature; then a second singularity, the successive decarbonations; corresponding to these singularities are traits of expression – not only the hardness, sharpness, and finish, but also the undulations or designs traced by the crystallization and resulting from the internal organization of the cast steel. The iron sword is associated with entirely different singularities, because it is forged and not cast or molded, quenched and not air cooled, produced by the piece and not in number; its traits of expression are necessarily very different because it pierces rather than hews, attacks from the front rather than from the side... [12]

In this quote Deleuze and Guattari do not make a clear distinction between properties (weight, color, sharpness) and capacities: the different ways in which a weapon can cut flesh or attack an enemy. The reason may be that both properties and capacities can express the identity of a weapon, that is, they are both traits of expression. But as noted earlier in this essay, there is an important distinction that must be made here: properties are always actual whereas capacities can be real but not actual, if they are not currently being exercised. And similarly for tendencies, which are also real even when not actually manifested, and which are always actualized as events. The difference between the saber and the sword, as assemblages, can then be established on the basis of a constellation of tendencies, capacities, and properties, or of singularities and traits of expression. As Deleuze and Guattari put it: "We will call an assemblage every constellation of singularities and traits deducted from the flow – selected, organized, stratified – in such a way as to converge ... artificially or naturally." [13]

With this definition we can take the concept of assemblage to very heart of matter, that is, we do not have to accept the requirement that all assemblages be "collective assemblages of enunciation", only that they all have expressive elements, whatever these may be. The crystalline sound of metals, their shine or gleam, but also the way they express what they can do as they exercise their electrical, mechanical, and chemical capacities: *metallic affects*. Chemically, metals are the second most powerful catalysts on the planet, losing only to biological enzymes. A catalyst is a molecular assemblage that can intervene in reality, to increase or decrease the speed of a chemical reaction, without itself being changed in the process. Electrically, metals are highly conductive, and are used by animals in atomic (or ionic) form to animate their brains and other parts of their nervous systems. Bringing assemblage theory down to the molecular or even atomic level promises to inject new life into materialist philosophy, a genre that used to include only a small set of material entities: physical labor; the food, drink, clothing and shelter needed to reproduce a work force; the means of production. These entities are, of course, still very important in the new materialism, but now *matter itself matters*. To conclude with an eloquent quote from Deleuze and Guattari:

> In short, what metal and metallurgy bring to light is a life proper to matter, a vital state of matter as such, a material vitalism that doubtless exists everywhere but is ordinarily hidden or covered, rendered unrecognizable, dissociated by the hylomorphic model. Metallurgy is the consciousness or thought of the matter-flow, and metal the correlate of that consciousness. As expressed in pan-metallism, metal is coextensive to the whole of matter, and the whole of matter to metallurgy. Even the waters, the grasses and varieties of wood, the animals, are populated by salts or mineral elements. Not everything is metal, but metal is everywhere. [14]

REFERENCES:

1. Gilles Deleuze and Felix Guattari. A Thousand Plateaus. (New York: University of Minnesota Press, 1987), p. 404.

2. Manuel DeLanda. War in the Age of Intelligent Machines. (New York: Zone Books, 1991), p. 62-79.

3. Gilles Deleuze and Felix Guattari. A Thousand Plateaus. Op. Cit. p. 397-398.

4. Ibid. p. 398.

5. Gilles Deleuze and Claire Parnet. Dialogues II. (New York: Columbia University Press, 2002), p. 69.

6. Gilles Deleuze and Felix Guattari. A Thousand Plateaus. Op. Cit. p. 88.

7. Ian Stewart and Martin Golubitsky. Fearful Symmetry. (Oxford: Blackwell, 1992), p. 104-109.

8. Gilles Deleuze and Felix Guattari. A Thousand Plateaus. Op. Cit. p. 381. (Italics in the original).

9. Thomas A. McMahon and John Tyler Bonner. On Size and Life. (New York: Scientific American Library, 1983), p. 155-162.

10. Gilles Deleuze and Felix Guattari. A Thousand Plateaus. Op. Cit. p. 402.

11. Ibid. p. 362-364.

12. Ibid. p. 408.

13. Ibid. p. 405-406. (Italics in the original).

14. Ibid. p. 406.

15. Ibid. p. 411.

Materialist Metaphysics.

Idealists have it easy. Their reality is uniformly populated by appearances or phenomena, structured by linguistic representations or social conventions, so they can feel safe to engage in metaphysical speculation knowing that the contents of their world have been settled in advance. Realists, on the other hand, are committed to assert the autonomy of reality from the human mind, but then must struggle to define what inhabits that reality. Many religious people, for example, are realists about transcendent spaces and entities, like heaven and hell, angels and demons. But a materialist metaphysician can only be a realist about *immanent* entities, that is, entities that may not subsist without some connection to a material or energetic substratum. And while it may be simple for a materialist to get rid of angelic or demonic creatures, there are other forms of transcendence that are far more difficult to remove.

In particular, if material entities are to have an identity that does not depend on human consciousness, the existence and endurance of this identity must be explained. The traditional way of accounting for a stable identity is by postulating the existence of *essences*, transcendent entities that have been part of realism for more than two thousand years and that are therefore hard to eliminate. The most defensible version of this concept is the one due to Aristotle. He defined metaphysics or ontology, as a science concerning itself with the study of entities capable of separate subsistence, entities about which the most important distinction was that between *those that subsist according to accident and those that subsist essentially.* [1] Metaphysics could not speculate about the accidental so it was the second kind of entities that constituted its subject matter. As he wrote:

Now, if there is something that is eternal and immovable, and that involves a separate subsistence, it is evident that it is the province of the speculative, that is, of the ontological, to investigate such. It is not, certainly, the province of the physical science, at any rate (for physical science is conversant about certain movable natures), nor of the mathematical, but of a science prior to both of these, that is, the science of metaphysics. For physical science, I admit, is conversant about things that are inseparable, to be sure, but not immovable; and of mathematical science some are conversant about entities that are immovable, it is true, yet, perhaps, not separable, but subsisting as in matter. But Metaphysics, or the First Philosophy, is conversant about entities which both have a separate existence and are immovable; and it is necessary that causes should be eternal, all without exception... [2]

Aristotle's world was populated by three categories of entities: *genus, species, and individual.* Entities belonging to the first two categories subsisted essentially, those belonging to the third one only accidentally. The genus could be, for example, Animal, the species Human, and the individual this or that particular person characterized by contingent properties: being white, being musical, being just. A series of subdivisions, in which at every step only logically necessary distinctions were made, linked a genus and its various species. Starting with the genus Animal, for example, we could first subdivide it into two-footed and many-footed types; then we could subdivide each type into differences of foot: hooves, as in horses, or feet, as in humans. When this series of subdivisions reached a point at which any further distinctions were accidental, like a foot missing a toe, we arrived at the level of the species, the lowest ontological level at which we could speak of an essence or of the very nature of a thing. As Aristotle summarized his realist ontology:

Physical or natural substances are acknowledged to have a subsistence; for example, fire, earth, water, air, and the rest of simple bodies; in the next place, plants, and the parts of these; animals, also, and their parts; and lastly; the heavens and the parts of the heaven... But, unquestionably from the foregoing reasonings the consequence ensues of there being other substances – I mean, the essence or very nature of a thing... Further, in other respects the genus is substance in preference to the species, and the universal to the singular. [3]

Aristotle is without doubt the most influential realist philosopher of all time. His ontological distinctions are today embedded in ordinary language, as when we say that a property is more generic or more specific than another. Replacing his metaphysics with something entirely different is, therefore, a major philosophical challenge. From the work of the philosopher

Gilles Deleuze we can derive such a novel ontology, an approach to the problem of existence that may be called a "neo-materialist metaphysics". In this approach all actual entities are considered to be *individual singularities*, that is, all belong to the lowest level of Aristotle's ontological hierarchy, while the roles of the two upper levels are performed by *universal singularities*. Later in this essay we will see that this is in fact only a rough characterization, since in some cases, such as that of humans or horses, the level of the species is replaced by an individual singularity operating at a larger scale. But for the purpose of establishing a sharp contrast to get the discussion started it will suffice to say that the Aristotelian categories of the general and the particular are replaced in a Deleuzian ontology by the universal singular and the individual singular.

The terms "general" and "universal" are often used interchangeably (by Aristotle or Deleuze) and they are also near synonyms in ordinary language, so the distinction between them must be a matter of technical definition. For Aristotle the levels of genus and species are directly linked to the logical role of predication, so that when we say, for example, that "Socrates is human" the proposition derives its truth from the fact that the particular individual named Socrates belongs to the general category "human", or what amounts to the same thing, that we can truly ascribe the general predicate "human" to the particular subject Socrates. On the other hand, the term "universal", in the technical sense used here, does not refer to logical predicates but to the *mathematical structure of a space of possibilities*. To summarize the main distinction between the two stances in Deleuze's own words we may say that "singularity is beyond particular propositions no less than universality is beyond general propositions." [4]

Let's begin the comparison of the two ontologies at the atomic scale, that is, with the case in which the genus is "Atom" and the species is "Hydrogen" or "Oxygen". A modern Aristotelian approach would begin by giving necessary and sufficient conditions to belong to the general category "Hydrogen", such as possession of a single proton (and a single electron). This is a perfectly reasonable way to specify the identity of this chemical species given that if we added another proton to a hydrogen atom we would change its identity, transforming it into an atom of helium. But in Aristotle a species did not just play a role in classifying entities but also in *generating* them. As a good realist, Aristotle knew that he had to explain how objective entities come into existence, in both nature and art. In both cases his explanation involved *essences acting as formal causes*. In nature, Aristotle saw the operation of essences as self-evident, from the observation that a horse begets a horse, and a human a human. In other words, he explained how animal species generate individual organisms by saying that they formally caused them. And similarly for art: in the case of building a house (or nurturing a patient to health) the formal cause is the idea that preexists in the human soul.

Hence, Aristotle argued that a house, or any other entity that "involves matter arises, or is generated, from that which does not involve a connection with matter: for the medicinal and the house-building arts are the form, the one of health, and the other of a house. Now, I mean by substance not involving any connection with matter, the essence or very nature or formal cause of a thing." [5] This is a much stronger claim than simply saying that possession of a single proton and a single electron is the criterion to belong to the category "Hydrogen". It is also a claim about what is philosophically significant about the generation of form: the process through which a house is built or a horse embryologically developed involves a connection with matter (immanent) and is therefore not as important metaphysically as the formal essence that is not so connected (transcendent).

In a Deleuzian ontology, on the other hand, an essence operating as a formal cause would not be what defines the identity of an *assemblage* composed of protons and electrons, nor would an essence make questions of *processes of assembly* irrelevant to metaphysics. The minimal definition of the term "assemblage" is that of a whole with properties that are both irreducible and immanent. An assemblage's properties are irreducible because while they emerge from the actual interaction between its parts, they cannot be ascribed to any of its parts. And they are immanent because if the components of the assemblage ceased to interact its own properties would cease to exist: emergent properties may not depend on this or that particular interaction, on this or that connection with matter, but they do demand that there should be some connection with matter. The emergent chemical properties (and capacities) of an atom, for example, depend on its outermost shell of electrons: whether the shell is missing an electron, or has an extra electron, or is exactly full. This property determines how many bonds an atom can form with other atoms: carbon atoms can form four; oxygen ones two; and hydrogen atoms only one. The properties of the outer shell (and the bonding capacities with which these endow an atom) are clearly not reducible to the properties of individual electrons, but they would cease to exist if those electrons stopped interacting with the atom's nucleus.

We can summarize this by saying that there is no such thing as "hydrogen in general", only a very large population of individual hydrogen atoms defined by properties that emerge from the continuous interaction among individual components. In other words, each hydrogen atom is an individual singularity. To the objection that even if each hydrogen atom is a unique historical entity all hydrogen atoms are basically the same (they are all defined by a one-proton nucleus) we can answer that there are other components, neutrons, that produce *inherent variation*. Depending on the number of neutrons a hydrogen nucleus possesses variant isotopes of this chemical species are generated: protium, deuterium, and tritium. The number of neutrons in a nucleus has very little effect on an atom's chemical properties, but it does affect its physical stability: some isotopes are stable and more enduring, while others decay faster. When we consider not one atom but an entire population of atoms, the relative abundances of isotopes, or more exactly, the statistical form of the distribution of isotopic variation, contains information about the historical processes that produced the members of the population, processes that replace formal causes in this ontology. In other words, the variation is not a trivial side effect but a significant source of knowledge.

Let's briefly sketch what is known in astrophysics about the production of atoms of different species. Although hydrogen and helium were produced under the intense conditions following the Big Bang, the rest of the chemical species had to wait hundreds of millions of years until the formation of stars. Today the nuclei of most atoms are assembled in stars, so the process of assembly is known as *stellar nucleosynthesis.* Stars of different sizes serve as assembly factories for atoms of different species: the larger and hotter the star the heavier the atoms it can put together. The smaller stars, like our Sun, are only hot enough (10 million degrees Kelvin) to burn hydrogen as fuel and produce helium as a product. At higher temperatures (over 100 million degrees), helium itself can be burnt as fuel and yield as products carbon, oxygen, and nitrogen. At even higher intensities (a billion degrees) carbon and oxygen become the fuel, while the products are atoms of the species: sodium, magnesium, silicon and sulfur. As intensities continue to increase silicon is burned as fuel to produce iron, and finally a maximum of intensity is reached in the process of explosive nucleosynthesis, in which the heavier species are created during the violent events known as "supernovae". [6]

We can imagine that, confronted with this information, Aristotle would be unimpressed, since he could argue that the details of how a house is built, or a patient healed, or an atom assembled, are less important than their formal causes. In particular, he could argue that regardless of what happens in stars, only a certain number of atomic species exists, a number that can be considered to have existed prior to any process of nucleosynthesis. There is, in fact, some truth to this objection which is why we need to add to an ontology of individual singularities the universal singularities that structure the space of possible species. Let's first consider this space as given in the famous Periodic Table of the Elements. The table itself has a colorful history because several scientists had already discerned regularities in the properties of the chemical species (when ordered by atomic weight) prior to Mendelev stamping his name on the table in 1869. Several decades earlier, for example, one scientist had already seen a simple arithmetical relation between triads of elements, and later on others noticed that certain properties (like chemical reactivity) recurred every seventh or eighth element. In other words, rhythms or periodically recurrent regularities had been observed pointing to the existence of a deeper structure. What constitutes Mendelev's great achievement is that he was the first one to have the courage to leave *open gaps* in the table instead of trying to impose an artificial closure on it. This matters because in the 1860's only around sixty species had been isolated, so the holes in Mendelev's table were like daring predictions that yet undiscovered species had to exist. He predicted, for example, the existence of germanium on the basis of a gap near silicon. The Curies later on predicted the existence of radium on the basis of its neighbor barium. [7] These risky predictions, and their eventual corroboration, is what gave the table its objective status. But what accounts for the underlying rhythms at the chemical heart of matter?

Before giving an answer let's take a quick look at the fields of mathematics that are relevant to the study of universal singularities. One of them is the study of differential equations, a field known today as "dynamical systems theory", and another is the field known as "group theory", a field that was born from the study of algebraic equations. The ancestor of the theory of dynamical systems is a mathematical method invented by the great eighteenth century mathematician Leonard Euler, "the calculus of variations", a method that could reveal the singularities structuring the space of possible solutions to differential equations. The singularities discovered by Euler were of a very simple type: minimum and maximum points. But the behavior of many physical systems is governed by minima or maxima of some quantity. The spherical shape of a soap bubble, for example, emerges spontaneously and recurrently because the entire population of molecules constituting a piece of soap film has the tendency to be in whatever state minimizes surface tension. The cubic shape of a crystal of ordinary table salt also emerges spontaneously and recurrently because its constituent atoms of sodium and chlorine have a tendency to minimize bonding energy.

In both cases, the space of possibilities contains a singularity, a topological point, that is real but need not be actual, if it is not currently being manifested. And when the singularity is actualized it leads to the formation of a variety of geometrical forms: spheres, cubes, and many other forms. This *divergent actualization* is the reason why mathematical singularities are referred to as "universal" and not "general": a general essence resembles that into which it becomes incarnated, but a universal singularity bears no resemblance to its divergent actualizations. [8] The discovery of the first singularities, simple as they were, was enough to provoke in Euler and his contemporaries the sense that they had revealed something about the divine plan, for how else, they thought, would a rational god organize his creation but by making the most efficient use of all materials, that is, by minimizing or maximizing.

Today we do not take such theological musings seriously but Euler's other insights are still valid. In particular, he thought of his discovery in Aristotelian terminology, calling singularities "final causes", because they represent long term tendencies, that is, the "final end" towards which a process tends. And these final causes, Euler argued, did not replace the study of mechanisms, that is, of processes involving efficient causes, but rather complemented it. [9] In other words, explaining the emergence of a bubble or a crystal involves both elucidating the different mechanisms that produce these forms (efficient causes) as well as determining the *mechanism-independent* tendencies common to both forms (final causes).

Euler's powerful insights on the structure of sets of possibilities were given an explicit spatial expression towards the end of the nineteenth century by another great mathematician, Henri Poincaré. Poincaré created the notion of *phase space* to study the space of possible solutions to nonlinear differential equations, and discovered new types of singularities as recurrent features of these spaces: different types of point singularities (steady state attractors); line singularities in the form of closed loops (periodic attractors); and he even glimpsed the existence of fractal singularities (chaotic attractors). [10] Like the original singularities, these different attractors represented the long-term tendencies of a process: the tendency towards a steady state; the tendency towards a simple rhythmic state; and the tendency towards a complex but stable rhythmic state. Although the ideas of Poincaré took decades to propagate outside of mathematics, by the 1960's they were in the air in cities like Paris. Gilles Deleuze was not only able to immediately grasp their significance but also possessed enough technical background to quickly adapt them for the formulation of metaphysical problems. In particular, he realized that universal singularities structure not only formal possibility spaces (spaces of possible solutions for equations) but also the possibility spaces associated with real entities, like bubbles or crystals. Deleuze also realized that there were other mathematical fields beside the differential calculus that could be used to reveal such structure, fields like group theory. [11]

Group theory was born from the study of the space of possible solutions to algebraic equations, but it eventually grew into an autonomous discipline concerned with the study of symmetry. Let's illustrate this using a soap bubble and a salt crystal. The salt crystal has the form of a cube, a figure that remains invariant if we rotate it by 0, 90, 180, or 270 degrees, in the sense that if an audience did not witness the performance of the rotation they would not notice that there has been any change. The sphere formed by the bubble, on the other hand, remains invariant under a much larger number of rotations: 0, 1, 2, 3 ... 359 degrees. In group theory this is expressed by saying that a sphere has more rotational symmetry than a cube. [12] A related concept is that of a *symmetry-breaking transition*, a transformation that yields a figure with less symmetry. If we constrain a piece of soap film so that it cannot form a sphere it can nevertheless manifest its tendency to minimize surface tension by forming a saddle-shaped surface (a hyperbolic paraboloid) that has less symmetry. In other words, group theory can allow us to study not only the tendency to generate a particular form but also the tendency to generate a family of such forms, each with a decreasing degree of symmetry. An example of a cascade of broken symmetries generating a family of forms starts with a sphere that loses symmetry to become a two-lobed figure, that, in turn, loses more symmetry and becomes a four-lobed figure, that, finally becomes an even less symmetric six-lobed figure.

Armed with this terminology we can now confront the question of what would replace the genus "Atom" in a Deleuzian metaphysics. The structure of the genus, the way it subdivides into species, is given by the rhythms of the Periodic Table. The first rhythm to be noticed, as mentioned before, was that the emergent properties of atoms recurred every eight species. Later on, however, as more species were discovered, chemists realized that the rhythm was more complex than that: it repeated twice with a cycle of eight; then it repeated twice more with a cycle of eighteen; then twice more with a cycle of thirty two. Adding to this the "lone" simplest species, hydrogen and helium, the series becomes: 2, 8, 8, 18, 18, 32, 32. The explanation for this complex periodicity turned out to be a symmetry-breaking cascade in the shape of the "trajectories" with which an electron "orbits" the nucleus. Actually, electrons do not move along sharply defined trajectories, since they behave like waves, but rather inhabit a cloud or statistical distribution possessing a given spatial form: an *orbital*.

The sequence of broken symmetries structuring the space of possible orbital forms may be unfolded as one injects more and more energy into a basic hydrogen atom. The single electron of this atom inhabits an orbital with the form (and symmetry) of a sphere. Exciting this atom to the next level yields either a second larger spherical orbital, or one of three possible orbitals with a two-lobed symmetry (two-lobes with three different orientations). Injecting even more energy we reach a point at which the two-lobed orbital becomes a four-lobed one (with five different orientations) which in turn yields a six-lobed one as the excitation gets intense enough. In reality, this unfolding sequence does not occur to a hydrogen atom but to atoms with an increasing number of protons in their nuclei, boron being the first chemical species to use the a non-spherically symmetric orbital. [13] Coupling this series of electron orbitals of decreasing symmetry to the requirement that only two electrons of opposite spin may inhabit the same orbital (a requirement that can be expressed in terms of regions of phase space) [14] we can reproduce, and explain, the rhythms of the Periodic Table.

Let's summarize the argument so far. In a Deleuzian ontology there is no such thing as "atoms in general" only variable populations of individual atomic assemblages. The kind and number of some components of the assemblage (protons) is what ensures that some properties are shared by all atoms of a given species, while the kind and number of others (neutrons) give these properties a certain degree of variation. Some variants of the assemblage will be highly stable isotopes, like the isotope of helium possessing exactly two protons and two neutrons, while other variants will lack this property. Only isotopes that are very stable last long enough in the intense environment of a star to serve as a platform for the assembly of more complex nuclei. This means that while the different species of atomic assemblages are defined by the structure of the space of possible electron orbitals, the production pathways from one species to another within a star are determined by populations of stable isotopes, a stability derived from the possession of a singularity (a minimum of energy) in the space of possible proton-neutron interactions. Thus, an atomic assemblage has an actual part, the components that actually interact to yield emergent properties, and a *virtual* part, the universal singularities and symmetries that structure its associated possibility space. The term "virtual" refers to the ontological status of entities that are real but not actual, such as tendencies that are not actually manifested (or capacities that are not actually exercised). The virtual component of an assemblage is called its *diagram*.

The overall metaphysical picture that emerges from these considerations is quite different from the Aristotelian one. In the latter the world is already *segmented* by logical categories, some more specific, others more generic, pre-existing segments that act as formal causes to generate all the particular members of each category. For Deleuze, on the other hand, the world is first and foremost a *continuum of intensity* that becomes segmented into species only as certain tendencies are manifested and certain capacities exercised. In the case of atomic assemblages the intensive continuum is embodied in stars, balls of plasma possessing a minimal segmentation but not entirely undifferentiated, since they have a structure defined by differences of temperature, pressure, and density. The possible ways of segmenting this continuum are not given by a logical subdivision of a genus into species, but by a virtual structure that can be captured mathematically.

At this point, Aristotle could raise an objection: this neo-materialist metaphysics derives most of its intuitions from physics and mathematics, but the whole point of creating a metaphysical science was precisely to discover speculatively those aspects of the world that are prior to both the physical and the mathematical. To this we may respond that the use of results from physics and mathematics does not imply that metaphysics can be subordinated to either one, only that it should not be satisfied with apriori speculation. In a Deleuzian ontology the main philosophical task is to extract, from the results produced by physicists and mathematicians, a *metaphysical problem*: why is the production and maintenance of atomic identity the way it is?.

A careless physicist might, for example, pose this problem incorrectly, asserting that atoms are what they are because of the immutable laws of nature. But this would hardly take us beyond Aristotle since it simply replaces one set of formal causes (general categories) with another one (general laws). Worse yet, most physicists espouse a positivist ontology, that is, they are committed to assert the mind-independent existence only of that which is directly observable. This means that they would not refer to temperature or pressure, for example, as emergent intensive properties of stars but as quantities, that is, as the quantities that are directly observable when reading a thermometer or a barometer. Similarly, when using the term "law" positivists would not refer to the *immanent patterns of being and becoming* to which the term refers in a realist ontology. [15] Instead, they would use the term to refer to the equations used to model those patterns, because equations are directly observable when written on a blackboard or printed on a piece of paper. Clearly, a positivist ontology makes it impossible for physicists to correctly pose metaphysical problems of existence. And similarly for mathematicians espousing a platonist ontology.

But what about the objection that the content of science is constantly changing and that building a metaphysics on top of it is like building a house on a foundation of sand?. It is true that the details of the causal mechanisms involved in a particular production process may change when new empirical findings determine that they are wrong, or when they are improved upon. And similarly for the mathematical ideas used to conceptualize the mechanism-independent structure of possibility spaces: in the future new advances in mathematics may reveal novel connections between the relevant fields (dynamical systems theory, group theory) or even generate entirely new fields. These innovations may indeed provide better tools to conceptualize universal singularities, much as novel empirical discoveries may provide new resources to conceptualize the processes that produce individual singularities. Metaphysics must remain sensitive to these potential changes, and remain capable of profiting from them.

Nevertheless, precautions must be taken. We must make an an effort, for example, to use only scientific results that have withstood the test of time, not the empirical findings of cutting edge science. It would be unwise to speculate about the problem of existence using what we know today about esoteric entities like dark matter or cosmic strings. But who could seriously doubt the existence of hydrogen or oxygen atoms? Although all scientific results are supposed to be falsifiable, this only implies that no result should be established apriori, not that all facts are equally likely to be false. To falsify the assertion that hydrogen or oxygen atoms exist, for instance, we would have to simultaneously falsify hundreds of other statements, like the statement that a water molecule is composed of two hydrogen atoms and one oxygen atom. But can any realist philosopher seriously consider such a wholesale falsification possible? It is crucial not to let metaphysical speculation be constrained by particular theories of science, especially theories taking as their object of study the reified generality "Science". In a neo-materialist metaphysics there is no such thing as "science in general", only a population of individual scientific fields, a population that is not converging on a final truth but rather growing and diverging as it tracks a reality that is itself divergent.

These remarks should be kept in mind as we attempt to eliminate other transcendent entities. In particular, replacing Aristotle's own example of the genus "Animal" and the species "Horse" or "Human", involves using results from a variety of biological fields, some of which (paleontology, evolutionary biology) are older and better established than others (genetics, embryology). Let's begin with a description of the results from the safest fields. Today it is widely accepted that biological species are as singular, as unique, and as historically contingent as individual organisms: species are born when their gene pool is closed to external flows of genetic materials through *reproductive isolation*, and they die through *extinction*. In other words, they belong to the category of entities that Aristotle regarded as incapable of yielding metaphysical knowledge, since as he wrote "things that are subject to corruption and decay are obscure to those even that are in possession of scientific knowledge". [16] Similarly, the defining properties of a biological species are not necessary, as Aristotle would require. Reproductive isolation is a contingent achievement that varies by degree, so nothing guarantees that the identity of a biological species will endure for ever. And since the anatomical, physiological and behavioral properties of organisms of a given species are produced by historical processes that cannot be exactly duplicated, driving a species to extinction is like killing an individual organism, that is, eliminating an entity that can never return again.

Clearly, the evolutionary conception of species is quite different from the Aristotelian one: the relation between organisms and species is not one of membership in a general category but one of reproductively interacting parts composing an emergent whole. In other words, species are nothing but assemblages of organisms, having the same ontological status (individual singularities) but operating at a larger spatio-temporal scale. [17] A common history in which the ancestors of the organisms that compose a new species faced similar challenges from predators and parasites, scarce resources and climatic changes, is what produces the bodily resemblances that we use to classify them. Shared selection pressures homogenize their gene pool, allowing us to infer that these organisms have more genes in common with each other than with organisms of other species. But as in the case of protons and neutrons, we need to stress not only what stays the same but also what varies. Without a constant production of genetic differences by accidental mutations or sexual recombinations, selection pressures would have no raw materials to operate on: no low fitness variants to filter out, or high fitness variants to promote.

Let's now describe the intensive continuum that is progressively segmented as new species emerge. As in the case of stars the term "continuum" does not imply an absolute absence of segmentation: stars may not be segmented by chemical species, except those that they themselves produce and burn as fuel, but they certainly are segmented by sub-atomic particles. Similarly, when speaking of an *ecological continuum,* the claim cannot be that it was not originally segmented physically or chemically, only that it was unsegmented biologically. Roughly, the first discrete biological segment to emerge consisted of flat layers of motionless bacteria inhabiting the interface between ocean water and bottom sediment. The earliest bacteria fueled themselves by fermenting available minerals but they eventually evolved the capacity to tap into solar radiation through photosynthesis, greatly enriching the intensive continuum. The term "intensive" refers to physical properties like temperature or pressure that can form *gradients* containing useful energy. A gradient of temperature is simply the result of coupling hot and cold masses of air or water, while a gradient of pressure results form coupling high and low pressure masses. Chemical gradients, in turn, can be created by coupling materials with different Ph, some acid, some alkaline, or materials with different properties of reduction and oxidation.

These coupled differences in intensity have the ability to drive processes, an ability that had to be harnessed by the earliest living creatures if they were to survive, proliferate, and evolve. [18] And once primitive bacteria developed the means to tap into physical and chemical gradients, they themselves became a biological gradient (a concentration of biomass) for the ancestors of today's amoebas and paramecia, that is, for the unicellular organisms that preyed on them. These primordial food chains possessed only a few distinct segments but as they complexified a much more thorough segmentation became possible, as new ecological niches opened up and new species came into being to occupy those niches. Today, solar energy and mineral nutrients are encapsulated into the bodies of countless animals and plants of different species, a highly segmented condition when compared to that characterizing the earliest biosphere.

Having replaced Aristotelian species, species like "Horse" and "Human" defined by a set of necessary and timeless characteristics, the next move is to identify the structure of possibility spaces that replaces the genus "Animal". Unfortunately, unlike the fields of evolutionary biology and ecology that provide relatively solid results ripe for metaphysical speculation, the biological disciplines on which we must rely for this task are at the cutting edge of research. Their results are, therefore, much more likely to change in the near future. As one embryologist suggests, the relevant results are today closer to the ontological status of dark matter than that of hydrogen or oxygen. [19] This means that we have to break the rule of using only results that have endured the test of time. Metaphysically, what we need to conceive is, in the words of Deleuze and Guattari:

> ... a *single abstract Animal for all the assemblages that effectuate it.* ... [For] the vertebrate to become an Octopus or Cuttlefish, all it would have to do is fold itself in two fast enough to fuse the elements of the halves of its back together, then bring its pelvis up to the nape of neck and gather its limbs together into one of its extremities... [20]

In other words, we need to conceive of a *topological animal* that can be folded and stretched into the multitude of different animal species that populate the world. Deleuze and Guattari do not, of course, believe that these topological operations can be performed on adult animals: only the embryos of those animals are flexible enough to endure these transformations. Moreover, the topological or virtual animal must be capable not only of being divergently actualized into many different assemblages, but each actualization must be *inheritable*. This means that the possibility space we are searching for is not just one of possible animal forms but also of possible genetic combinations that can reliably and recurrently produce those forms. There is one more point that must be clarified before we begin to describe these possibility spaces. While in the Aristotelian classification there are only two levels, in modern taxonomies there are many levels, with species and genera at the bottom, and kingdoms and phyla at the top. Aristotle's "Animal" genus is today the animal kingdom (Animalia), subdivided into different phyla (Chordata, Arthropoda), in turn subdivided into sub-phyla like vertebrates and insects. Unlike the lower taxa, the higher ones are defined by a common *body plan,* the features of which are often defined early on in embryological development. It is the structure of the space of possible body plans that replaces the genus "Animal".

The first problem we must confront is that the possibility spaces involved are very different from the ones with which we have over a century of experience, like phase space and its topological singularities. Unlike phase space, which is fully continuous and has a well defined spatial structure, *the space of possible genes* is entirely discrete and has no intrinsic spatial order. In addition, what is singular, special, or remarkable (let alone universal) in the structure of discrete combinatorial spaces is not well understood. So we are in uncharted waters here and must proceed carefully. Let's begin with what we do know. The genetic code itself is by now well established and used routinely in industry to create proteins out of genes, or to create a designer gene from a natural protein. Both genes and proteins are linear sequences of molecules differing only in their components: nucleotide molecules in the case of genes, aminoacids in the case of proteins. The genetic code is simply a way of mapping one type of molecular sequence into another, three nucleotides corresponding to each aminoacid, the correspondence itself being arbitrary, a kind of frozen evolutionary accident.

The one thing we know for sure about the spaces of possible genes and proteins is their enormous size. The number of possible sequences of a given length is the number of available components raised to the maximum length. Genes are composed of only four possible nucleotides, while proteins can draw from a repertoire of twenty possible aminoacids. A very short protein five amino acids long can exist in over three million different combinations (the number twenty raised to the fifth power). The number of possible proteins three hundred amino acids long, the average length of a contemporary enzyme, is literally infinite, larger than the number of seconds the universe has existed. And similarly for genes. Although in this case the number of different components is much smaller, their length is three times greater since they must use three nucleotides to specify a single aminoacid. Thus, in either case we are considering combinatorial spaces that grow explosively as the length of the sequences increases. But how can we study the structure of these infinite spaces if they lack any intrinsic spatial order?.

One strategy would be to impose on them a non-arbitrary order, that is, an order that has some connection to the tendencies and capacities of the molecular sequence in question. In the case of genes the capacity to replicate is crucial, as is the tendency for copying errors (mutations) to occur during replication. So a reasonable spatial order can be imposed if we arrange each sequence of nucleotides so that it has as neighbors all sequences that differ from it by only one mutation. The resulting space is multidimensional because it must include all the variants that can be created by varying each nucleotide along the full length of a given gene, and the resulting one-mutant variant must be assigned a different dimension. But while the sheer number of dimensions makes the space very complex it also greatly simplifies the distribution of sequences: after we rearrange them every possible gene is *in direct contact with all its one-mutant neighbors,* forming a connected path that evolution can follow. In other words, given this spatial arrangement genetic evolution can be visualized as a continuous walk from one neighbor to the next, driven by events producing one mutation at a time. [21]

To capture the selection pressures that guide these walks we can superimpose on this combinatorial space a set of fitness values, one for each possible sequence, reflecting their contribution to the overall fitness of an organism. This yields a distribution of singularities, that is, a distribution of *local maxima and minima of fitness.* In rare cases there will be a single, easy to reach, global maximum so we can apply the old formula of "survival of the fittest". But in most realistic cases even if there is such a global maximum it may be surrounded by local maxima that can trap evolving species: once a peak of fitness has been climbed by an evolutionary walk, descending from it to climb a higher peak may be prevented by selection pressures, since any gene below the peak is by definition less fit. Genetic drift, on the other hand, may help species break away from a local trap by providing a random mechanism of variation not subjected to the filtering effects of selection.

Applying these ideas to the task of replacing the genus "Animal" involves distinguishing between different types of genes, since only relatively few genes contribute to the specification of body plans. The first distinction that must be made is between genes that perform routine housekeeping tasks on every cell of an organism and those that cause the differences between different cell types: bone, muscle, blood, nerve, skin. Given that all the cells that compose an organism have the exact same DNA there must be special genes that turn other genes on or off in different cell types giving them their distinct character. These are genes that code for proteins that have DNA itself as their target, binding to a portion of it to determine whether another gene "downstream" will or will not be expressed. This type of gene can itself be further differentiated into those that control nearby downstream genes, and those that, in addition, are controlled by upstream genes. Producing a protein that switches other genes on or off, while simultaneously being capable of being switched on or off, gives these genes the ability to form circuits and networks of switches. If we consider that the central processing unit of a desktop computer is just such a network of switches (And-gates, Or-gates, Not-gates) the power of this type of genes becomes obvious.

Including only genes that can form part of such networks in the possibility space greatly diminishes its size but such spaces can still be quite large. The switches themselves are non-coding regions of DNA to which proteins attach and are typically between six and nine nucleotides long: this gives us between 4096 (4^6) and 262,144 (4^9) possible permutations. The genes producing the proteins that perform the switching are relatively few in number, but if we assume that only five hundred of the twenty thousand coding genes in the human genome are involved in embryological development, that gives us two hundred and fifty thousand possibilities for two-gene circuits; over twelve million possibilities for circuits of three genes; and over six billion possibilities for circuits of four genes. [22] We may further restrict the size of the possibility space by concentrating only on those genes for which there is evidence of direct involvement in body plan specification. These are the so-called *Hox genes.*

Several characteristics set these special or singular genes apart: they are extremely old, predating the Cambrian explosion (over five hundred million years ago) the fossils of which give us the firmest evidence about the divergence of body plans; they are clustered together in the animal genome and their spatial arrangement has intriguing correspondences with the distribution of body parts that characterizes a body plan; and they display striking similarities across phyla. Vertebrates, for example, have four Hox clusters (thirty nine genes) while insects have two Hox clusters (eight genes.) [23] Thus, the possibility space that we would need to explore is the space of all combinations (circuits, networks) formed by the set of Hox genes affecting the *early stages* of development of an embryo. Superimposed on this combinatorial space there would have to be, as in the previous example, a set of fitness values creating a distribution of singularities. In this case, however, the selection pressures determining the fitness values should not be external, like predators or parasites, but internal: existing circuits and networks that select mutations preserving their coherence, and select against those that disturb it. [24] The picture that emerges from these considerations is one of a set of genetic networks defining the possibilities open to a given animal phylum, possibilities that are actualized in the long term by ecological factors and in the short term by embryological ones. As Deleuze writes:

How does actualization occur in things themselves?.... Beneath the actual qualities and extensities [of things themselves] there are spatio-temporal dynamisms. They must be surveyed in every domain, even though they are ordinarily hidden by the constituted qualities and extensities. Embryology shows that the division of the egg is secondary in relation to more significant morphogenetic movements: the augmentation of free surfaces, stretching of cellular layers, invagination by folding, regional displacement of groups. A whole kinematics of the egg appears which implies a dynamic. [25]

To understand how Hox genes (and other switchable genes operating during embryogenesis) transform a fertilized egg into a horse or a human, to stick to Aristotle's examples, it will be useful to describe what is known about this process. The easiest way to visualize the process is to relate certain recurrent features of the adult form to those in the embryo at different stages of development. The first set of correspondences is between the symmetries and broken symmetries (or polarities) of the adult form, and the axes that, like longitude, latitude, and altitude, define the *geography of the embryo*. Adult vertebrates have bilateral symmetry, right and left sides being roughly invariant if mirror imaged, but head and tail, as well as face and back (top and bottom in horses), break this symmetry. The population of cells that constitutes an early embryo develops an East-West and a North-South axis, corresponding to these two broken symmetries, by activating certain genes on stripes of cells. In effect, what used to be a continuous cellular population becomes segmented along longitude and latitude, at progressively finer scales, and then the future identity of these segments is established by the activation of Hox genes within their cellular sub-populations. [26]

The second set of correspondences is between the modular construction of the adult form and a modular use of Hox genes. Animals are assemblages of components, many of which are repeated modules differing only in kind and size. For example, limbs are made of parts (thigh, calf, ankle; upper arm, forearm, wrist) and their extremities (feet and hands in humans, hooves in horses) are also made out of variably repeated modules. Similarly, all vertebrates possess a stiff vertebral column but the number and type of vertebrae varies from one species to the next. [27] A limb begins its actualization as a small bud that projects out of the embryo at a specific location along the East-West axis. Then the growing bud is itself segmented by its own sets of local longitudes and latitudes, each segment containing cellular sub-populations in which specific genes are switched on. The extremities of these limbs are, in turn, further segmented through a set of finer subdivisions, prefiguring the future digits. [28]

To this production of modular *expression patterns* of Hox genes – concentration gradients of Hox gene products in cellular sub-populations at specific latitudes and longitudes – we must add further spatio-temporal dynamisms defining collective migrations of cells, internal differentiation of cells into bone and muscle (and other cell types), foldings and stretchings creating closed pockets and invaginations. [29] For our purposes here this sketch should suffice since, as I remarked above, the next few decades are likely to produce many new discoveries, as well as novel approaches to the mathematical modeling of genetic circuits and networks, so parts of the description above may have to be modified. From a metaphysical point of view, however, what matters is that the relevant causal agents (Hox genes, genes marking axes of longitude and latitude, cellular populations) do not act as formal causes but as efficient causes, and this conclusion is unlikely to be modified in the future even as we refine our knowledge of the causal mechanisms. To summarize: Aristotelian species like "Horse" and "Human" should be replaced by historically constituted species that have the same ontological status as the organisms that compose them, that is, that are individual singularities; and the genus "Animal" should be replaced by a space of possibilities in which the different body plans are universal singularities, capable of being divergently actualized into a large number of sub-phyla and classes.

To finish this essay let's add more detail to the philosophical concept that is at the basis of the replacement strategy: the concept of assemblage. Both atoms and animals are assemblages of building blocks that vary in kind and number. The use of the term "kind" here is, of course, problematic, since a kind is simply a category. But the difficulty here is more apparent than real: the building blocks used as components of an assemblage are themselves assemblages operating at a smaller scale, and we should be able to give causal mechanisms defining the processes that actualized them, as well as the mechanism-independent structure of their own possibility spaces. In other words, we are always dealing with assemblages of assemblages, the part-to-whole relation recurring at different scales. Moreover, the intensive continua (the stars, the coupled system hydrosphere-atmosphere, the early embryo) within which chemical and biological assemblages are actualized are themselves assemblages, albeit relatively unsegmented ones. Gilles Deleuze, in fact, uses the term "assemblage" only for this latter type of entity: atoms and animals would not be strictly speaking assemblages but what he refers to as *strata*. What then justifies using the term "assemblage" indiscriminately to refer both to the intensive continua and to the segmented forms that are born in these continua?.

The fact that introducing terms that define categories of entities creates problems for our strategy, since assemblages and strata could be considered to be species of a genus. A way out of this dilemma would be to use a single term, the term "assemblage", but build into it a parameter quantifying the degree to which an entity is segmented or unsegmented. We can use the term "territorialization" as a name for this parameter, a territory being simply an area of an environment that has been divided or segmented. This way a "stratum" becomes an entity in which the territorialization parameter has a high value, and an assemblage (in its Deleuzian meaning) an entity with a low value for this parameter. To the objection that this approach reduces what in Deleuze is a qualitative difference to a quantitative one, we can respond that many parameters are characterized by critical points of intensity marking transitions from quantity to quality, such as the phase transitions between gases, liquids, and solids. The territorialization parameter would have to be conceived in a similar way, with strata being not a separate kind of entity but simply a *different phase* in which assemblages can exist. The opposition between strata and assemblages would then be:

> entirely relative. Just as milieus swing between a stratum state and a movement of destratification, assemblages swing between a territorial closure that tends to restratify them and a deterritorializing movement that connects them to the Cosmos. Thus it is not surprising that *the distinction we were seeking is not between assemblage and something else, but between two limits of any possible assemblage.* [30]

If chemical and biological species exemplify one of these two limits, the stratified or fully segmented extreme, what is the other limit? I introduced above the term "diagram" to refer to the structure of the possibility space associated with any actual assemblage, and I said that the ontological status of diagrams is designated by the term "virtual", that is, something that is real but not actual. The assemblage of all virtual diagrams is the other extreme state, the properly cosmic limit referred to in the quote. The existence of this limit state is another crucial difference between the two metaphysical pictures we have been comparing: while the Aristotelian ontology has several distinct levels, defined by the hierarchy genus-species-individual, the Deleuzian ontology is flat: the world of actual assemblages forming a *plane of reference*, that is, a world of individual singularities operating at different spatio-temporal scales, to which we can refer by giving them, for example, a proper name; and the world of virtual diagrams defined by universal singularities forming *a plane of immanence,* a plane that does not exist above the other plane (like a genus that is ontologically "above" a species) but is like its reverse side. A single flat ontology with two sides, one side populated by virtual problems and the other by a divergent set of actual solutions to those problems.

The non-hierarchical nature of this ontology not only exorcises transcendence – all transcendent entities need higher levels or dimensions in which to subsist – but it also makes it possible to relate the two planes dynamically. One aspect of this dynamic has already been mentioned: unlike an Aristotelian world in which everything is pre-segmented by logical categories, in this ontology the production and maintenance of the segments themselves must be explained by processes of territorialization. Metaphysically, these processes can be conceived as movements following the direction going from the plane of immanence to the plane of reference. But movements in the other direction are also possible. The territorialization of atoms or animals takes place within assemblages, intensive continua, that are *relatively deterritorialized,* while the plane of immanence itself is the result of a movement that is *absolutely deterritorialized,* yielding an ideally continuous space defined exclusively in terms of topological invariants (dimensionality, connectivity, distribution of singularities). Using a parametrized concept facilitates thinking about this double dynamic, since all three different assemblage states can be conceived as phases in which quantity change into quality: "The absolute expresses nothing transcendent or undifferentiated. It does not even express a quantity that would exceed all given (relative) quantities. It expresses only a type of movement qualitatively different from relative movement." [31]

Thus, all actual assemblages are subject to historical processes that territorialize them, rigidifying their segments, or that deterritorialize them, making their segments more supple or even fully continuous. This implies that what is deterritorialized is not just what comes before what is territorialized, like a fertilized egg that comes before the fully formed animal body. This may be true for an individual organism but not for the species: at any given time the reproductive communities making up a given species will contain many adult bodies but also many embryos at different stages of segmentation. Moreover, most multicellular species are forced to go back to the unsegmented, unicellular stage if they are to evolve, because it is mutations in the genes that control embryological development that are most important in generating morphological variation. In other words, multicellular species are forced to deterritorialize every generation prior to territorializing again. In this sense, the biological egg is not what comes before and is left behind. And similarly for other intensive continua, like stars conceived as cosmic eggs. As Deleuze and Guattari write:

But the egg is not regressive; on the contrary, it is perfectly contemporary, you always carry it with you as you own milieu of experimentation ... The egg is the milieu of pure intensity, spatium not extension, Zero intensity as a principle of production. There is a fundamental convergence between science and myth, embryology and mythology, the biological egg and the psychic or cosmic egg: the egg always designates this intensive reality, which is not undifferentiated, but is where things and organs are distinguished solely by gradients, migrations, zones of proximity. [32]

What does it mean to say that we carry an "egg" with us as a milieu of experimentation? It is a reference to the fact that relative deterritorializations can be achieved not only by gaining continuity or losing segmentation but also by exercising capacities to affect and be affected. Whereas properties fix the identity of a segment, capacities can allow one segment to interact with an entirely different one forming a new assemblage within which that fixed identity may undergo a metamorphosis. A human body may be definable by a finite set of properties, some extensive (height, weight) others intensive (blood pressure, body temperature), but is is also defined by capacities to perform a potentially infinite set of activities. These include not only productive activities whose proliferation is attested by the progressive differentiation of labor into many specialities (blacksmiths, carpenters, potters, soldiers) but also by unproductive activities that nevertheless express the body's infinite potential: jugglers, tight-rope walkers, trapeze artists. Both sets of activities insert the human body into an assemblage, an "egg", within which deterritorializations and new territorializations take place.

As a realist philosopher, Aristotle recognized the existence of capacities; the fact that unexercised capacities are real but not actual; and even the distinction between capacities to affect and capacities to be affected. Thus, he spoke of potentialities in both humans and things, some of which are passive and others active: the capacity of fire to warm another body or the capacity of a human being to build a house, but also the capacity of fat to be burned or of flesh to be bruised. [33] But within his pre-segmented ontology the dynamic of segments exercising capacities in interaction with other segments to produce movements of territorialization and deterritorialization cannot be formulated.

A powerful illustration of this dynamic can be given by going back to the simplest of chemical segments: the humble hydrogen atom. As mentioned before, all atoms have certain emergent properties like having an outer shell that has an excess or deficit of electrons or, on the contrary, that is completely "full". This property, in turn, defines the capacities of atoms to bond with other atoms: if the shell is full, as in the noble gases, this capacity will be very low (noble gases form only a few compounds) while a deficit or excess translates into a much greater capacity to form bonds and compounds. Most atoms have the capacity to form *covalent* bonds – extremely strong bonds formed by the sharing of a pair of outer shell electrons – with other atoms. This capacity is, in this sense, quite ordinary, although the result of its actual exercise is crucial for the maintenance of the identity of material entities, since covalent bonds are the glue that holds molecules, and things made out of molecules, together. But in addition to this ordinary bonding capacity, hydrogen atoms possess a singular or special capacity to form weaker bonds, appropriately called "hydrogen bonds".

This other ability can only be exercised in very special circumstances defined both by capacities to affect and be affected: the atom, or group of atoms, that are the target of the bonding operation must be electronegative, and the hydrogen atom itself must be covalently attached to another atom, or group of atoms, that is also electronegative. The fact that hydrogen bonds are easier to form and break makes them less "territorializing" than covalent bonds, that is, the molecular assemblages they form are less rigidly articulated than those formed by covalent bonds. An example of the deterritorialized molecular assemblages made possible by hydrogen bonds are genes and proteins: while the identity of a particular gene is maintained through time by its covalent bonds, its capacity to self-replicate is defined by hydrogen bonds, since the two strands of the double helix must be easily unglued, and new nucleotides easily glued to each strand serving as a template. [34]

Therefore, in a very real sense, it is the existence of a singular bonding capacity, one not shared by all atoms, that permits the existence of the singular capacities at the basis of all living creatures. To put this in the terminology of assemblage theory, the existence of a less territorializing bond is instrumental in bringing about the powerful deterritorialization that characterizes self-replicating molecules (genes), as well as molecules that can recognize targets and accelerate chemical reactions (enzymes). Or to put this even more metaphysically: a deterritorialization at an atomic scale yields another deterritorialization at the scale of biological macromolecules, the two movements forming a line of flight carrying matter away in the direction of the plane of immanence.

Although Deleuze and Guattari make only a passing reference to the territorializing effect of different kinds of chemical bond, they clearly recognize the importance of the discoveries of modern genetics for a materialist metaphysics. As they write:

The alignment of the code or linearity of the nucleic sequence in fact marks a threshold of deterritorialization of the "sign" that gives it a new ability to be copied and makes the organism more deterritorialized than a crystal: only something deterritorialized is capable of reproducing itself. ... It is the crystal subjugation to three-dimensionality, in other words, its index of territoriality, that makes the structure incapable of formally reproducing and expressing itself ... On the contrary, the detachment of a pure line of expression on the organic stratum makes it possible for the organism to attain a much higher threshold of deterritorialization, gives it a mechanism of reproduction covering all the details of its complex spatial structure, and enables it to put all its interior layers "topologically in contact" with the exterior, or rather, with the polarized limit (hence the special role of the living membrane). [35]

For similar reasons they value discoveries from fields like ecology and ethology, to perform the properly metaphysical task of tracking the deterritorializations in living matter that have led to the progressive differentiation of the biosphere. Ecological relations, the relations between a predator and its prey, for example, involve both the properties defining the actual states of an organism (the state of being hungry) as well as the exercise of their capacities: the ability to hunt, the ability to evade a hunter. Capacities need not be actual, if not currently exercised, and when they do become actual it is not as states but as events that are always double: to eat-to be eaten. When capacities are exercised over many generations they can lead to extended series of interactions in which living creatures of different species force each other to deterritorialize: predators and their prey can enter into genetic "arms races", in which any inheritable improvement in the ability to evade predators or capture prey acts as a selection pressure for the development of countermeasures. In effect, through this mutual stimulation, predators and their prey can force each other to adaptively modify their genetic identity: a deterritorialization followed by a reterritorialization. Symbiosis can also lead to series of deterritorializations and reterritorializations, as Deleuze and Guattari note in the case of plants and the insects that pollinate them:

The orchid deterritorializes by forming an image, a tracing of a wasp; but the wasp reterritorializes on that image. The wasp in nevertheless deterritorialized, becoming a piece in the orchid's reproductive apparatus. But it reterritorializes the orchid by carrying its pollen. ... a becoming-wasp of the orchid and a becoming-orchid of the wasp. Each of these becomings brings about the deterritorialization of one term and the reterritorialization of the other; the two becomings interlink and form relays in a circulation of intensities pushing the deterritorialization ever further. [36]

Perhaps the most striking example of the deterritorializing effect of symbiosis occurred in some of the most ancient organisms, causing one of the earliest evolutionary divergences. As mentioned above, ancient bacteria managed to discover all the major ways to tap into physical and chemical gradients (fermentation, photosynthesis, respiration), each discovery drastically increasing the amount of captured energy and pushing organisms further from thermodynamic equilibrium. The intensification of the energy flow meant that bacterial populations did not just reproduce their numbers but produced a net surplus, creating a concentration gradient formed by their flesh as it accumulated, opening up opportunities for the differentiation of ecological relations. In other words, a biological gradient was formed and with it the possibility of using it as an energy source by creatures that could prey on bacteria. [37] The earliest predators, on the other hand, did not reinvent the machinery behind the three basic energy extraction strategies: they simply internalized the bodies of bacteria as so many building blocks, entering with them into an intimate form of symbiosis called *endosymbiosis*. [38] Even today, the descendants of those internalized creatures live within us, as mitochondria in animals and chloroplasts in plants, proving that the two symbionts have shared a common line of deterritorialization (as well as mutual reterritorializations).

This line can be seen as a continuation and intensification of the one in which the emergence of genetic information marked a threshold of deterritorialization of the sign (information). But if DNA allowed organisms to deterritorialize over many generations it also represented a new way of rigidifying biological segments, not by territorialization this time but by coding. As an assemblage, an animal may be said to be highly coded if its behavior is rigidly determined by its genes, and relatively decoded if it can learn during its life time. In the case of pollinating insects, for example, most of the behavior is genetically "hard-wired" but they nevertheless exhibit a certain degree of decoding, because they can learn through Pavlovian conditioning to associate the presence of flower nectar with colors, odors, and multi-petalled shapes. [39] We can capture this novel condition of biological assemblages by adding to the concept   a second parameter, one quantifying the *degree of coding and decoding*. That is, we can conceptualize the historical specialization of genetic information into a separate causal factor by thinking of assemblages, up to this point possessing a single parameter, as suddenly acquiring a second one. And as the proliferating populations of neurons began to wrestle control of behavior from the genes, the intensive quantity measured by the second parameter also crossed thresholds, including the thresholds of deterritorialization and decoding that produced us, humans, as a separate species endowed with very special capacities, like bipedal locomotion and complex manual skills. To conclude with the words of Deleuze and Guattari:

> Not only is the hand a deterritorialized front paw; the hand thus freed is itself deterritorialized in relation to the grasping and locomotive hand of the monkey... [There are also] correlative deterritorializations of the milieu: the steppe as an associated milieu more deterritorialized than the forest, exerting a selective pressure of deterritorialization upon the body and technology (it was on the steppe, not on the forest, that the hand was able to appear as free form, and fire as a technologically formable matter). Finally, complementary reterritorializations must be taken into account (the foot as a compensatory reterritorialization for the hand, also occurring on the steppe). Maps should be made of all these things, organic, ecological, and technological, maps one can layout on [the plane of immanence.] [40]

REFERENCES:

1. Aristotle. The Metaphysics. (New York: Prometheus Books, 1991), p. 100.

2. Ibid. p. 124-125.

3. Ibid. p. 167.

4. Gilles Deleuze. Difference and Repetition. (New York: Columbia University Press, 1994), p. 163.

5. Aristotle. The Metaphysics. Op. Cit. p. 142.

6. Stephen F. Mason. Chemical Evolution. (Oxford: Clarendon Press, 1992), Chapter 5.

7. P. W. Atkins. The Periodic Kingdom. (New York : Basic Books, 1995), Chapter 7. p. 72-73.

8. Gilles Deleuze. Difference and Repetition. Op. Cit. p. 212

        "Actualization breaks with resemblance as a process no less than it does with identity as a principle. In this sense, actualization or differenciation is always a genuine creation."

9. Leonard Euler. Quoted in Stephen P. Timoshenko. History of Strength of Materials. (New York: Dover, 1983), p. 31.

        "Since the fabric of the universe is most perfect, and is the work of a most wise Creator, nothing whatsoever takes place in the universe in which some relation of maximum and minimum does not appear. Wherefore there is absolutely no doubt that every effect in the universe can be explained as satisfactorily from final causes, by the aid of the method of maxima and minima, as it can from the effective causes themselves.... Therefore, two methods for studying effects in nature are open to us, one by means of effective causes, which is commonly called the direct method, the other by means of final causes. ... One ought to make a special effort to see that both ways of approach to the solution of the problem be laid open; for thus is not only one solution greatly strengthened by the other, but, more than that, from the agreement of the two solutions we secure the highest satisfaction."

10. June Barrow-Green. Poincare and the Three Body Problem. (Providence: American Mathematical Society, 1997), p. 32-33.

        See also:

    Ian Stewart. Does God Play Dice: The Mathematics of Chaos. (Oxford: Basil Blackwell, 1989), p. 70-71.

11. Gilles Deleuze. Difference and Repetition. Op. Cit. p. 179-180.

12. Joe Rosen. Symmetry in Science. (New York: Springer-Verlag, 1995), Chapter 2.

13. Vincent Icke. The Force of Symmetry. (Cambridge: Cambridge University Press, 1995), p. 150-162.

14. Stephen F. Mason. Chemical Evolution. Op. Cit. p. 60.

15. Mario Bunge. Causality and Modern Science (New York: Dover, 1979), p. 22-23.

16. Aristotle. The Metaphysics. Op. Cit. p. 160.

17. Michael T. Ghiselin. Metaphysics and the Origin of Species. (Albany: State University of New York Press, 1997), p. 78.

18. Ronald E. Fox. Energy and the Evolution of Life. (New York: W.H. Freeman, 1988), p. 58-59.

19. Sean B. Carroll. Endless Forms Most Beautiful. The New Science of Evo Devo. (New York: W. W. Norton, 2005), p. 113.

20. Gilles Deleuze and Felix Guattari. A Thousand Plateaus. (New York: University of Minnesota Press, 1987), p. 255.

21. Manfred Eigen. Steps Towards Life. (Oxford: Oxford University Press, 1992), p. 92-95.

22. Sean B. Carroll. Endless Forms Most Beautiful. Op. Cit. p. 118-119.

23. Wallace Arthur. The Origin of Animal Body Plans. A Study in Evolutionary Developmental Biology. (Cambridge: Cambridge University Press, 2000). p. 156-157.

24. Ibid. p. 222.

25. Gilles Deleuze. Difference and Repetition. Op. Cit. p. 214.

26. Sean B. Carroll. Endless Forms Most Beautiful. Op. Cit. p. 92-95.

27. Ibid. p. 20-21.

28. Ibid. p. 102-104.

29. Wallace Arthur. The Origin of Animal Body Plans. Op. Cit. p. 97-98.

30. Gilles Deleuze and Felix Guattari. A Thousand Plateaus. Op. Cit. p. 337. (My italics.)

31. Ibid. p. 509.

32. Ibid.p. 164.

33. Aristotle. The Metaphysics. Op. Cit. p. 178-179.

34. Jean-Marie Lehn and Philip Ball. Supramolecular Chemistry. In The New Chemistry. Edited by Nina Ball. (Cambridge: Cambridge University Press, 2000), p. 302.

35. Gilles Deleuze and Felix Guattari. A Thousand Plateaus. Op. Cit. p. 59-60.

36. Ibid. p. 10.

37. George Wald. The Origin of Life. In The Chemical Basis of Life. (San Francisco: W.H. Freeman, 1973), p. 16-17.

38. Jan Sapp. Living Together: Symbiosis and Cytoplasmic Inheritance. In Symbiosis as a Source of Evolutionary Innovation. Edited by Lynn Margulis and Rene Fester. (Cambridge: MIT Press, 1991), p. 16-17.

39. James L. Gould. Ethological and Comparative Perspectives on Honey Bee Learning. In Insect Learning. Edited by Daniel R. Papaj and Alcinda C. Lewis. (New York: Chapman and Hall, 1993), p. 31-38.

40. Gilles Deleuze and Felix Guattari. A Thousand Plateaus. Op. Cit. p. 61.

Intensive and Extensive Cartography.

As biological organisms and as social agents we live our lives within spaces delimited by natural and artificial *extensive boundaries*, that is, within zones that extend in space up to a limit marked by a frontier. Whether we are talking about the frontiers of a country, a city, a neighborhood, or an ecosystem; or about the defining boundaries of our own bodies – our skin, our organ's outer surfaces, the membranes of our cells– inhabiting these extensive spaces is part of what defines our social and biological identities. We also inhabit other spaces, *zones of intensity,* the boundaries of which are not defined by spatial limits but by critical thresholds: the zones of high pressure explored by deep-sea divers; the zones of low gravity lived by astronauts; the zones of low temperature experienced by arctic explorers; the zones of high speed traversed by test pilots. These are all, of course, rare professions, but we all populate these intensive zones even if at much moderate intensities.

Extensive and intensive spaces can both be mapped, but the maps will necessarily be different. An extensive map captures features of the Earth that are extended in space, such as coastlines, mountain ranges, or the areas of land and volumes of air space defining the sphere of sovereignty of a given country. By contrast, an intensive map captures differences in the intensity of a particular property (gradients) as well as the dynamic phenomena that are driven by such gradients. A well-known example, appearing on our television screens every night, is a meteorological map showing zones of high and low pressure, cold and warm fronts, air masses moving slowly or rapidly. Although the distinction between extensive and intensive properties is old, dating back to medieval scholastic philosophy, in its modern form it has been developed mostly by physicists. So we can begin our discussion with the textbook definition:

Thermodynamic properties can be divided into two general classes, namely intensive and extensive properties. If a quantity of matter in a given state is divided into two equal parts, each part will have the same value of intensive properties as the original, and half the value of the extensive properties. Pressure, temperature, and density are examples of intensive properties. Mass and total volume are examples of extensive properties. [1]

A typical extensive property, such as length, area, or volume, is divisible in a simple way: dividing an area into two equal parts results in two areas with half the extension. But if we take a volume of water at, say, ninety degrees of temperature, and divide it into two half volumes, we do not get as a result two parts having forty five degrees of temperature each, but two parts with the same original temperature. Put differently, while two extensive quantities add up in a simple way, two pieces of land adding up to a proportionally larger piece of land, intensive quantities do not add up but rather average: two volumes of water or air at different degrees of temperature, when placed into contact, trigger a diffusion process that tends to equalize the two temperatures at some intermediate value. Gilles Deleuze is the only modern philosopher who grasped the importance of this distinction, not only adopting the textbook definition but extending it to highlight its metaphysical significance. In particular, Deleuze established a *genetic* relation between the extensive and the intensive: the diversity of entities that we can perceive directly are entities bounded in extension, but they are generated by invisible processes governed by differences of intensity.

A good example is the diversity of entities that populate the atmosphere: hurricanes, thunder storms, cloud formations, wind currents. These entities inhabit our consciousness as meteorological phenomena but we can't normally perceive the gradients of temperature, pressure, or speed that are responsible for their genesis. Similarly, while many diverse animals appear to us as entities bounded by their skin, we are not normally aware of the gradients of concentration of gene products, the biochemical differences of intensity, that generate those animals through an embryological process. In short, the diversity that is given to us in phenomenological experience depends for its existence on what is not so given. Or as Deleuze puts it:

Difference is not diversity. Diversity is given, but difference is that by which the given is given. ... Difference is not phenomenon but the nuomenon closest to the phenomenon. ... Every phenomenon refers to an inequality by which it is conditioned. Every diversity and every change refers to a difference which is its sufficient reason. Everything which happens and everything which appears is correlated with orders of differences: differences of level, temperature, pressure, tension, potential, differences of intensity. [2]

Thus, although in its modern form the distinction between intensive and extensive properties belongs to thermodynamics, it is in the context of a materialist philosophy that it acquires its properly metaphysical significance. A similar point applies to extensive and intensive maps. Let's discuss in some detail these two types of map, concentrating first on their scientific aspects, then extracting the relevant metaphysical problems they pose. Since the time of Ptolemy map makers have struggled with the problem of capturing into a flat representation the spherical features of our planet. One could, of course, simply use a globe, a spherical map, in which the spatial relations can be represented directly. But if the goal is to create a flat map that can be folded and carried around, the spherical form of our planet must be transformed somehow, because spheres are not the kind of shapes that can be unrolled and made to lie flat. Cylinders and cones, on the other hand, are just those kind of shapes, so if one could transform a sphere into a cylindrical or conic shape then the problem would be solved. The special transformation that achieves this objective is called "projection".

Ptolemy projected the sphere into a cone, much as one would project a slide into a screen, while Mercator, fourteen hundred years later, used a cylinder as his screen. Although once unfolded and flattened both the conic and the cylindrical representations give the desired result, a new problem emerges: one can preserve the original spatial relations yielding specific shapes, like the shape of a coastline or a mountain range, or one can preserve the original areas covered by land or water masses, *but not both*. Opting for the former (in what is called a "conformal" map) we lose the true relations between areas or between lengths, while choosing the latter (an "equal-area" map) gives us shapes that appear distorted in the map. For the purposes of navigation along a coastline, where visual recognition of landmark shapes is what matters, a conformal map is the right choice, but for statistical purposes, to depict the density of population per square mile, for example, we need an equal-area map. Other uses call for a compromise, a projection that does not preserve anything unchanged, but in which the errors are small enough or balance each other out. [3]

From a metaphysical point of view there are only two things that matter in this brief description: the existence of transformations – two in this case, a projection operation corresponding to shining light on a piece of film, and a section operation, the equivalent of intercepting those light rays on a screen – and the fact that once applied, these transformations leave some of the features of the original form invariant. In the traditional Mercator projection, for instance, shapes remain invariant, as do some lengths (the distances along the line of the equator) but areas and distances away from the equator do not remain unchanged. These two concepts, *transformations and invariants*, would be destined to play an increasingly important role in science, eventually becoming an integral part of twentieth-century physics. A good example of the impact that these two concepts had in mathematics is the change in status that Euclidean geometry suffered in the nineteenth century.

In the late eighteenth century most philosophers and scientists agreed that Euclidean geometry was not only the most fundamental of all geometries but the one that captured the features of real physical space. This privileged status did not change when mathematicians invented other geometries that were not flat but curved, but that also were, like their more prestigious relative, *metric* geometries, geometries in which the length of a line or the angle between two lines are fundamental concepts. But then came a momentous change as mathematicians realized that all metric geometries were in fact a special case of projective geometry, that is, that the basic metric concepts (length, angle, shape) could be logically derived from the non-metric concepts of projective geometry. [4] Thus, what up to that point had been a humble geometry belonging to minor fields of science, fields like cartography, became the most fundamental.

The shift in logical priority among the geometries took place when mathematicians realized that transformations could be brought together into special sets called "groups", and that invariants relative to these groups could be used to define geometric entities. A cube, for example, can be characterized by a list of its properties, that is, by the fact that it has six sides, or that each side is a square. But it can also be defined by how it is affected, or not affected, by certain transformations, by the fact that, for example, its visual appearance remains unaffected after undergoing rotations of 90°, or multiples of 90°, on any axis. These rotations can be grouped together forming the set {90°, 180°, 270°, 360°}. A sphere, on the other hand, remains invariant under rotations by any amount of degrees, so its group is much larger: {1°, 2°, 3°, 4° ... 360°}. In the theory of groups an entity with properties that remain invariant under a larger group of transformations is said to have *more symmetry* than one with a smaller group. In this example, a sphere has a higher degree of symmetry than a cube under rotational transformations, and this fact can be used to classify it as a geometrical figure. [5]

Felix Klein, one of the most important mathematicians of the nineteenth century, realized that this idea applied not only to individual geometrical figures, but to the different geometries themselves. Metric geometries, Euclidian and non-Euclidean, form spaces the properties of which remain invariant by a group containing rotations, translations, and reflections. In other words, lengths, angles, and shapes remain invariant under this group of *rigid* transformations. In projective geometries, on the other hand, those properties do not remain invariant but others do, such as linearity, collinearity, and the property of being a conic section. Moreover, the group of transformations that leave the latter invariant is a larger set, including rotations, translations, and reflections, but also projections and sections. It was this realization, that *the group characterizing metric spaces is a subgroup of the one characterizing projective spaces*, that established the logical priority of the latter. [6] When geometries like topology were invented the followers of Klein realized that topological spaces had invariants under even larger groups, including transformations like stretching and folding. This led to the idea of classifying all the known geometries by their degree of symmetry: topology had the most symmetry followed by differential geometry, projective geometry, affine geometry, and Euclidian geometry. Mathematicians view this classification as a purely logical construction, but it is possible to extract from it a metaphysical lesson by making the relations between the different spaces genetic. In this metaphysical version metric spaces would be literally born from non-metric ones as the latter progressively lose symmetry (or break symmetry) and gain invariants. Euclidian geometry, far from being the most basic of all geometries, would be a byproduct of a more fundamental geometry, topology, produced as the latter undergoes a cascade of *symmetry-breaking events.*

Other aspects of Klein's classification lend themselves to a metaphysical treatment. As we move down the cascade, for example, more and more figures become distinct. Whereas in Euclidean geometry small and large circles, small and large ellipses, small and large parabolas, are all different figures, at the next level up (affine geometry) circles, ellipses, and parabolas of all sizes are the same; and one more level up (projective geometry) all conic sections are one and the same figure. This is explained by the fact that if a figure can be transformed into another using only transformations in the group then the two figures are the same. Affine geometry has in its group the scaling transformation, so the size of a given conic section is not relevant to establish its identity. Similarly, in projective geometry tilting the screen on which a circle is projected transforms it into an ellipse, and moving the screen so that part of the ellipse is now outside of it yields a parabola. In other words, all conic sections are inter-convertible using transformations in the projective group, so they are all one and the same figure. [7] The folding and stretching transformations available in topology take us beyond this: all closed figures (triangles, squares, pentagons, circles) are inter-transformable so they are all the same. Metaphysically, this suggests that a symmetry-breaking cascade represents *a process of progressive differentiation,* that is, a process that takes relatively undifferentiated topological figures and through successive broken symmetries generates all the different metric figures.

Having extracted the significant metaphysical problems, genetic problems, from the geometrical spaces related to extensive maps, let's do the same for intensive maps. What need to be mapped in this case are not the borders of entities possessing a spatial organization, like the boundaries of an ocean, a lake, or another body of water, but *thresholds of intensity causing spontaneous transformations* in the spatial organization of those bodies. These transformations are called "phase transitions". Imagine a frozen body of water, a solid piece of ice, linked to an outside supply of energy that we can control. As we increase the amount of energy flowing into the system its temperature reaches a critical point at which, suddenly, the ice begins to melt. At that intensive threshold a solid spontaneously changes into a liquid as its spatial organization, its manner of occupying space, mutates. If we continue to increase the amount of energy we reach another critical threshold, the boiling point of water, and the liquid turns into a gas, with accompanying changes in extensive properties: the amount of space the water molecules occupy, their volume, greatly expands. Finally, as the temperature reaches yet another threshold, first the molecules of water dissociate into their component atoms, then even the atoms of hydrogen and oxygen lose their own identity, the entire population becoming an electrified cloud of charged particles: a plasma.

A map of these intensive thresholds is called a *phase diagram*. The number of dimensions of the map is determined by the number of intensive parameters used to affect the body of water. Using a single parameter, temperature, yields a map that is one-dimensional, that is, the temperature values form a linear series in which the thresholds appear as points: the point at zero degrees centigrade marking the melting point of water, and the one at one hundred degrees centigrade marking its boiling point. (The names of the points vary depending on the direction in which the thresholds are crossed: in the opposite direction they are the freezing and condensation points respectively.) These two singular points are constant – so constant that we use them to mark our thermometers – but only as long as we keep other possible parameters unchanged. In particular, zero and one hundred degrees mark thresholds at sea level altitude, but the precise numerical value changes if we are on a tall mountain because the pressure that the air exerts diminishes with altitude. This implies that adding a second parameter, pressure, changes the map into a two-dimensional space in which the thresholds cease to be points to become lines. And similarly when we add a third parameter, like specific volume: now the thresholds are surfaces in a three-dimensional map.

Typically, as we add more dimensions an intensive map reveals further complexity in the behavior of matter. The two-dimensional phase diagram of water, for example, is not structured by two parallel lines running through the zero and one hundred degrees points of temperature. If it were, it would not add any extra information to what we already have in the one dimensional case. Rather, the lines form a shape with the form of the letter "Y". At sea level pressure the map is structured by the upper part of the Y, so a perpendicular line of temperature values intersects its two arms at the two points just mentioned. But at lower pressures the map is shaped by the lower part of the Y, so a line of temperature values intersects it only once. This means that at very low pressures, such as there exist in outer space, there are only two distinct phases, solid and gas, one transforming directly into the other in a phase transition called "sublimation". Finally, despite the fact that the thresholds are now lines, singular points are also present: the point in the Y where the two upper arms meet the lower vertical is called a "triple point", a *zone of coexistence* at which all three phases simultaneously occur and can be readily transformed into one another. Similarly, the right arm of the upper Y does not cross the entire map but terminates at a critical point creating a *zone of indiscernibility* within which the liquid and gas phases of water become indistinguishable. [8]

Let's give a different example of an intensive map, one using speed as a control parameter. The behavior of fluids in motion exhibits sudden changes in form at critical thresholds of speed, that is, it undergoes phase transitions between different *regimes of flow*: at low speeds the flow is uniform or steady (laminar); then past a threshold it becomes wavy or periodic (convective); and past yet another threshold it becomes turbulent, displaying a fractal structure of eddies within eddies. An intensive map of these transformations would be one-dimensional, a line of speed values divided into three different regimes by critical points. Using a special laboratory apparatus consisting of two transparent concentric cylinders between which a fluid is sandwiched, we can enrich this one-dimensional map by spinning the inner cylinder and carefully studying the effect of speed on the fluid. The higher degree of control allowed by the so-called Coutte-Taylor apparatus reveals seven distinct regimes of flow: laminar, Taylor vortex flow, wavy vortex flow, modulated wavy vortices, wavy turbulence, turbulent Taylor vortices, and featureless turbulence. If we modify the apparatus so that we can spin both the inner and outer cylinders we can create a two dimensional map. As before, adding an extra dimension reveals much hidden complexity. Two new intensive zones are created on both sides of the line with its seven regimes of flow: to the right there are variations produced when the two cylinders spin in the same direction, including ripples, twisted vortices, corkscrew wavelets; to the left there are variations produced by spinning them in opposite directions, such as simple spirals, interpenetrating spirals, and spiral turbulence. [9]

These two intensive maps capture the possible spatial organizations of water (solid-liquid-gas) or its possible ways of flowing (laminar-convective-turbulent), while the critical thresholds mark points at which a quantitative change, an extra degree of temperature or speed, becomes a qualitative change. There are two lessons here for metaphysics. The first lesson is that, as Deleuze argues, intensive properties are not so much indivisible, as *that which cannot be divided without changing nature*. [10] The thresholds do segment an intensive map but each subdivision corresponds to a different variant phase or regime. The second lesson is that the critical thresholds are always one dimension lower than the map itself. If, like mathematicians, we use the variable "n" to indicate the number of dimensions, we can say that *intensive thresholds always have n-1 dimensions*: points in a line, lines in a surface, surfaces in a volume. The reason why this is significant is that in a materialist metaphysics the structure of possibility spaces must always be immanent not transcendent, and as Deleuze argues, transcendent forms of determination always exist on a higher dimension than the space in which a material process unfolds. That is, transcendent determination is always n+1. Aristotelian essences, for example, exist on a higher ontological plane than that of the individual entities they formally determine, the level of species or genus, endowing these individuals with homogeneity and unity from above. The immanent structure of possibility spaces, on the other hand, "however many dimensions it may have, ... never has a supplementary dimension to that which transpires upon it. This alone makes it natural and immanent." [11] The two lessons from intensive maps add up to *a metaphysics of immanent variation* that replaces a metaphysics based on transcendent unity. Or what amounts to the same thing, a metaphysics in which the Multiple replaces the One:

> The multiple *must be made*, not by always adding a higher dimension, but rather in the simplest of ways, by dint of sobriety, with the number of dimensions one already has available – always n-1 (the only way the one belongs to the multiple: always subtracted). Subtract the unique from the multiplicity to be constituted; write at n-1 dimensions. [12]

Intensive maps can be enriched by "drawing" them on topological surfaces instead of on a flat Euclidean plane. The maps just described (phase diagrams) are merely graphic representations of laboratory data: parameters are carefully varied in a controlled setting and for each combination of values the resulting phase or regime of flow is recorded; the entire set of values is then given graphic form to display the thresholds and the zones of stability they demarcate. But ideally, an intensive map should have inherent features, such as a certain distribution of singularities that are invariant and that correspond to those stability zones. To understand how these "enhanced" intensive maps can be created, and the further metaphysical insights they may yield, let's return for a moment to the history of geometry. In the early decades of the nineteenth century when mathematicians wanted to study a space like a curved two-dimensional surface, they used the old Cartesian method: they embedded the surface into a three-dimensional space (a space with one supplementary dimension, that is, n+1) structured by a set of axes; then, using those axes, they assigned coordinates to every point of the surface. In this way the surface became a set of x, y, and z coordinates, the relations between which were expressed using algebra.

But then an entirely new approach was created by tapping into the resources of the differential calculus. In particular, if relations between the changes of two or more quantities could be expressed as a rate of change, then the calculus allowed finding the instantaneous value for that rate. If the changing quantities were spatial position and time, for instance, the instantaneous value for the rate of change of one relative to the other, that is, the instantaneous velocity, could be computed. Applying this idea to geometry involved thinking about a curved surface as an object characterized by the rate at which its curvature changes between two points, and then using the calculus to compute "instantaneous" values for this rate of change: a surface ceased to be a set of coordinate values and became *a field of rapidities and slownesses*, the rapidity or slowness with which curvature changed at each point. Friedrich Gauss was the first to realize that the calculus could be used to study a surface *without any reference to a global embedding space*, that is, using only local information on the surface itself. That is, Gauss "advanced the totally new concept that a surface is a space in itself". [13]

Gauss solved the two dimensional case while his disciple, Bernhard Riemann, whom everyone expected to tackle the three dimensional case, went ahead and solved the n-dimensional one. An n-dimensional space defined using the calculus is called a "differential manifold" or a "multiplicity". Eventually, multiplicities acquired more symmetry, that is, they were given properties that remained invariant under a larger group of transformations, and became topological. Is it possible to "draw" intensive maps on multiplicities?. To do this we need to include in the map not only the parameters characterizing the environment of the entity to be studied, but also the variables defining the entity itself. Each different variable quantifies one way in which the entity is free to change, that is, one degree of freedom. Then, we must assign each degree of freedom to one of the dimensions of the multiplicity. This generates a geometrical representation of *the space of possible states* in which the object can be, each state characterized by a particular combination of values for its degrees of freedom.

This space is called "state space" (or "phase space") and was invented by the mathematician Henri Poincaré towards the end of the nineteenth century. The physical entity's state at any instant becomes a point in state space, while the behavior the entity displays as it changes states become a trajectory (a series of points). As he studied these possibility spaces Poincaré noticed that trajectories tended to converge at special points in the space, as if they were being attracted to them: it did not matter where the trajectory had its origin, or how it wound its way around the space, its long term tendency was to end up at a singularity. [14] These special or singular points were eventually named *attractors*. When a state space has several attractors, its singularities are surrounded by an area within which they affect trajectories, an area called a "basin of attraction": if a trajectory begins within a particular basin of attraction then it inevitable ends up at the attractor. This implies that attractors and their basins define zones of stability, since they pin down trajectories to a particular set of properties (combinations of the values for the degrees of freedom) and do not let them escape.

This, and the fact that attractors are topological invariants, remaining unchanged no matter how we deform the space, suggests that these enhanced intensive maps capture objective features of the entity under study. In particular the trajectories in the map have long term tendencies that correspond to the long term tendencies of the physical entity being mapped, a fact that can be confirmed using laboratory experiments. Finally, parameters can be added as a set of control knobs with hash marks indicating the critical thresholds: as we vary a parameter and a threshold is reached a *bifurcation* occurs, an abrupt change in which one distribution of attractors is transformed into another, topologically inequivalent, one. These bifurcations correspond to transitions between phases or regimes of flow. Hence, unlike phase diagrams that only create a graphic record of the results of laboratory measurements, the maps constituted by state spaces posses zones of stability, and display transitions between zones, *as inherent features*, yielding a more intimate relation between a map and that which is being mapped.

Gilles Deleuze quickly realized the metaphysical importance of these ideas. To begin with, the fact that a multiplicity is defined locally without the need for a global embedding space implies that we are dealing with immanent features of a space, not with extrinsic, transcendent coordinates defined through the use of a supplementary dimension. When Deleuze uses the term "multiplicity", moreover, is not to refer to a manifold as an abstract geometric object, but to a manifold the dimensions of which have already been assigned an intensive property. In other words, the term "multiplicity" in Deleuze refers to manifolds used to conceptualize possibility spaces. Concepts defined not linguistically but as the structure of a space of possibilities, like the space of possible colors, he designates as "Ideas". As he writes:

> An Idea is an n-dimensional, continuous, defined multiplicity. Color -or rather, the Idea of color - is a three dimensional multiplicity. By dimensions, we mean the variables ... upon which a phenomenon depends; by continuity, we mean the set of relations between changes in these variables – for example a quadratic form of the differentials of the [variables]; by definition, we mean the elements reciprocally determined by these relations, elements which cannot change unless the multiplicity changes its order and its metric. When and under what conditions should we speak of a multiplicity? There are three conditions which together allow us to define the moment when an Idea emerges; 1) The elements of the multiplicity must have neither sensible form nor conceptual signification... They are not even actually existent, but inseparable from a potential or a virtuality. ... 2) These elements must in effect be determined, but reciprocally, by reciprocal relations that allow no independence whatsoever to subsist. ... In all cases the multiplicity is intrinsically defined, without external reference or recourse to a uniform space in which it would be submerged. ... 3) A multiple ideal connection, a differential relation, must be actualized in diverse spatio-temporal relationships, at the same time that its elements are actually incarnated in a variety of terms and forms. The Idea is thus defined as a structure. [15]

Multiplicities are *virtual*, that is, real but not actual, and capable of *divergent actualization*. The tendency of liquid water to become ice or steam, for example, is real at all times even if the water is not actually undergoing a phase transition. And the phase transitions themselves can be actualized in a large variety of materials in which the details of the condensation or crystallization mechanisms may be very different. This gives a definite ontological status to the topological invariants of a space that, in its scientific usage, may be considered only a means to explore the structure of mathematical models (the structure of the solutions to differential equations). This is precisely what the role of the philosopher should be: to extract the metaphysically significant concepts from what scientists and mathematicians produce as part of their own investigations, concepts that must be extracted without reintroducing reified generalities or essences. If we, for example, gave virtual multiplicities the ontological status of mere possibilities then, as modal logic and its theory of possible worlds has shown, we must bring essences back into the picture. [16] Unlike modal logicians Deleuze is not committed to assert the mind-independent existence of the possibilities themselves: possibilities are real but only when we entertain them by considering alternative scenarios. On the other hand, Deleuze certainly believed in the objective reality of the topological invariants (number of dimensions, distribution of singularities) that structure concrete possibility spaces:

> The virtual is not opposed to the real but to the actual. *The virtual is fully real in so far as it is virtual....* Indeed, the virtual must be defined as strictly a part of the real object – as though the object had one part of itself in the virtual into which it plunged as though into an objective dimension.... The reality of the virtual consists of the differential elements and relations along with the singular points which correspond to them. The reality of the virtual is structure. We must avoid giving the elements and relations that form a structure an actuality which they do not have, and withdrawing from them a reality which they have. [17]

After this journey through the world of extensive and intensive maps we have gathered a set of metaphysical insights that can be weaved into a world-view consisting of three separate but related domains. First, there is the domain of final products, defined by their extensive properties: the length, area, or volume of the space they occupy; the number of components they have; the amount of matter and energy they contain. Second, there is the domain of production processes, defined by intensive differences, the flows driven by these differences, and the critical thresholds that change quantity into quality. And third, there is the domain of virtual structure accounting in a purely immanent way for the regularities in the processes and the products. The intrinsic divisibility of extensive properties plays a key role in this world-view because the divisions or segmentations are constitutive of actual entities: mineral nutrients and solar energy can exist as gradient-driven flows defining an intensive ecological continuum, but also segmented or encapsulated into the bodies of plants and animals.

There is one more distinction from thermodynamics that we need to complete this picture: the distinction between *the molecular and the molar*. In the mid-nineteenth century, physicists took for granted the existence of intensive properties, like the temperature or pressure of a body of water or air, and concentrated on studying and modeling their relations: what happens to temperature as pressure increases, for example. But late in that century the problem of how those properties emerge from the interactions between water or air molecules became important, and gave rise to an entirely different field: statistical mechanics. At that point a distinction was made between the molar properties of an entire population of water or air molecules, and the molecular dynamics of the members of the population themselves. The relation between the molar and the molecular, therefore, corresponds to that between an emergent whole, or *assemblage,* and the parts that compose it. On the other hand, because what is an assemblage at one scale can be a component part at another scale, the names "molecular" and "molar" can be misleading. A thunderstorm, for example, is an assemblage of flows of air and water in different regimes of flow. In particular, flows forming circular patterns (convection cells) are like the "moving parts" of the storm. In this case, the term "molar" would apply to the entire thunderstorm while the term "molecular" would refer to the convection cells themselves, even though each cell is made up of millions of molecules. Like the distinction between micro and macro, the one between molecular and molar should be made relative to the part-to-whole relation.

We can borrow this terminology but, as before, we must modify it to adapt it to metaphysical speculation. In its philosophical sense the term "molar" refers to the rigid segments making up final products, defined not only by their extensive properties but also by their intensive properties *at equilibrium*, that is, by the molar properties (average temperature, average pressure) that emerge once a gradient has been cancelled. The term "molecular", in turn, refers to the smaller segments that are the component parts of a molar aggregate, but always taken as a dynamic population of interacting micro-segments, a population defined by intensive differences that are maintained through the continuous injection of a flow of matter or energy (a population *far from equilibrium*). It is only in this condition, a molecular population in which gradients are kept alive, that fluxes and thresholds define a process, not a product. If we studied the micro-segments by themselves, detaching them from their dynamic interactions, they would be molar entities regardless of their absolute scale.

We can summarize this by saying that the actual world is constituted by two separate but related segmentarities: one molar or rigid defining finished products, from atoms and molecules, to institutional organizations and cities; the other molecular or supple defining flows and thresholds that enter into the production and maintenance of the molar segments. When the processes behind the production and maintenance of identity are ignored, when philosophers concentrate on making lists of the extensive and intensive properties defining the final products, metaphysics becomes transcendent, leading to the creation of static typologies that use a product's properties to classify it, and then reify those defining properties into eternal essences. But when those processes are made into an indispensable component of a metaphysics *the taxonomic is replaced by the cartographic*, the philosopher using n-1 entities (lines on a plane) to create maps: molar lines of rigid segmentarity, molecular lines of supple segmentarity, and the lines of flight that connect the previous two to the virtual. As Deleuze puts it:

Whether we are individuals or groups, we are made up of lines and these lines are very varied in nature. The first kind of line which forms us is segmentary – of rigid segmentarity: family-profession; job-holiday; family-and then school-and then the army-and then the factory-and then retirement. ... In short, all kinds of clearly defined segments, in all kinds of directions, which cut us up in all senses, packets of segmentarized lines. At the same time, we have lines of segmentarity which are much more supple, as it were molecular. It is not that they are more intimate or personal, they run through society and groups as much as individuals. ... But rather than molar lines with segments, they are molecular fluxes with thresholds or quanta. ... Many things happen on this second line – becomings, micro-becomings, which don't even have the same rhythm as "our" history. ... At the same time, again, there is a third kind of line, which is even more strange: as if something carried us away, across our segments, but also across our thresholds, towards a destination that is unknown, not foreseeable, not pre-existent. ... the line of flight and of the greatest gradient... [18]

In his work with Guattari, Deleuze uses these three types of lines to create maps of social processes at different scales: the scale of the individual, the group, or the entire social field. What is mapped in each case is an assemblage, since "any assemblage necessarily includes lines of rigid and binary segmentarity, no less than molecular lines, or lines of border, of flight or slope." [19] At the largest scale there are assemblages of entire cultures, such as the assemblage that emerged in the European continent during the fall of the Roman Empire: the roman cities, their governmental organizations, their geometrically organized military camps and rigid phalanxes, are mapped with molar lines; the movements of the nomads from the steppes (the Huns), with their highly flexible and mobile armies, are mapped with lines of flight; while the migrant barbarian tribes caught in the middle, and pushed by the Huns against the empire, are assigned molecular lines. [20]

A different example, at a smaller scale, would be the assemblage of urban and rural settlements (and the organizations that exercise authority in those settlements) composing an archaic empire, like the Egyptian empire. In this case, their map shows that the semi-autonomous agricultural villages at the periphery of the empire possess a supple segmentarity, while the central state apparatus in its urban capital displays the most rigid form of segmentarity. [21] A line of flight in this case could be illustrated with a mobilized state army that, returning triumphant from a far away military victory, resists being demobilized, thereby threatening the very stability and identity of the state apparatus.

Each of the components of these assemblages must, in turn, be mapped. Thus, a bureaucratic organization – a single component of a state apparatus – with its rigidly segmented offices, schedules, task assignments, and written regulations, is also composed of personal and professional networks formed by its staff, networks that display a more supple segmentarity. [22] If we imagined a technological innovation, a new communication tool that, for example, allowed these networks to be mobilized to reform the organization, this may be mapped by a line of flight, that is, a series of events causing a change in the identity of the organization. At the smallest scale that is significant for social explanation, the scale at which the persons that staff a bureaucratic organization operate, we need maps of their bodies and minds, considered as assemblages or molar aggregates of sub-personal components: "Take aggregates of the perception or feeling type: their molar organization, their rigid segmentarity, does not preclude the existence of an entire world of unconscious micropercepts, unconscious affects, fine segmentations that grasp or experience different things, are distributed and operate differently." [23] The onset of a delirium, whether due to mental illness, a high fever, or the ingestion of psychedelics, can liberate those micropercepts and accelerate their escape from a molar subjectivity, changing in the process the identity of the person.

In all these cases we are dealing with *assemblages of assemblages*, each level of scale needing its own map. The term "level of scale", on the other hand, must always be used in a relative not an absolute sense. The human body, for example, is composed of large populations of individual cells, but also of large populations of individual atoms. Relative to the part-to-whole relation cells operate at a larger scale than atoms but in absolute terms the two populations are *coextensive*, that is, they both extend to the limits of the entire body. And similarly for larger social assemblages: "the two forms are not simply distinguished by size, as a small form and a large form; although it is true that the molecular works in detail and operates in small groups, this does not mean that it is any less coextensive with the entire social field than molar organization." [24]

Of the three kinds of lines used in these maps the hardest to conceptualize is the third one. If the first two lines are used to map the relations between actual products and actual processes, the third line maps the links between all actual segments, molar or molecular, and the unsegmented or continuous virtual world. This virtual dimension can also be conceptualized as an assemblage, one in which the components are multiplicities. Since multiplicities vary in their number of dimensions, each physical entity having a different number of degrees of freedom, the assemblage they form must be such that it accommodates this variability. This assemblage is referred to as the "plane of immanence" or "plane of consistency". As Deleuze and Guattari write:

> It is only in appearance that a plane of this kind 'reduces' the number of dimensions; for it gathers in all the dimensions to the extent that *flat multiplicities* –which nonetheless have *an increasing or decreasing number of dimensions*– are inscribed upon it. ... Far from reducing the multiplicities' number of dimensions to two, the *plane of consistency* cuts across them all, intersects them in order to bring into coexistence any number of multiplicities, with any number of dimensions. The plane of consistency is the intersection of all concrete forms... [25]

The symmetry-breaking cascade created by Felix Klein to organize the different geometries can be put to metaphysical use to visualize the connections between these three domains of reality. We can envision an ideally continuous space, the plane of immanence, that would progressively become discontinuous, first by becoming incarnated into intensive continua differentiated only by gradients, and in which the broken symmetries appear as critical thresholds, and then by becoming fully "metric", broken down into separate rigid segments: physical, chemical, biological, and social segments. We can summarize all this in the following metaphysical formula: *material reality is generated as a topological and intensive continuum progressively differentiates into extensive segments, as thresholds are crossed and symmetries broken.* Or more accurately, material reality emerges from a process of actualization that goes from the virtual to the intensive away from equilibrium, and from there to the extensive and the intensive at equilibrium.

In addition to this Deleuze postulates the existence of another process running in the opposite direction: *counter-actualization.* The need to postulate such a counter-process derives directly from the materialist restriction on transcendent entities and spaces. In particular, if this ontology is to include an assemblage of all the virtual multiplicities we need to give an account of the creation and maintenance of this immanent space. Else, it will be nothing but a Platonic heaven in which essences have ceased to be metric (the essence of "sphericity") to become topological. Thus, we need a mechanism for the production and reproduction of immanence. As Deleuze puts it:

> Many movements, *with a fragile and delicate mechanism*, intersect: that by means of which bodies, states of affairs, and mixtures, considered in their depth, succeed or fail in the production of ideal surfaces [plane of immanence]; and conversely, that by means of which the events of the surface are actualized in the present of bodies (in accordance with complex rules) by imprisoning their singularities within the limits of worlds, individuals, and persons. [26]

Actualization always takes place in the "present of bodies", that is, all actual events occur in the present time. But the production of "ideal surfaces" must take place in another temporality, one without any actually occurring events. Much as non-metric spaces, spaces in which lengths or areas are meaningless concepts, help us think about the spatial aspects of the plane of immanence, we must try to conceive of a non-metric time proper to it. If metric space is defined by rigid lengths that are measurable and divisible, *chronometric time* must be thought as defined by rigid durations that are the "lived" and measurable presents of actual entities, from the longest cosmic or geological presents, to the shortest sub-atomic ones. A topological form of time would, in turn, be one in which the notion of temporal duration is meaningless. Only singularities can be used to think about this non-chronometric time: the minimum thinkable continuous time and the maximum thinkable continuous time; a present without any duration whatsoever that is unlimitedly stretched in the past and future directions simultaneously, so that nothing ever actually happens but everything just happened and is about to happen. [27]

When speculating about actualization we can use raw materials from a diverse set of scientific and mathematical fields, but when investigating counter-actualization we are on our own. This is, therefore, the most properly metaphysical area of this ontology. Lines of flight belong here, as the components parts of mechanisms of immanence. More exactly, Deleuze and Guattari distinguish between *absolute* lines of flight that depart from the actual taking away with them only the subtlest components of events (topological invariants) as raw material for the plane of immanence; and *relative* lines of flight that follow this escape from the actual but only up to a point at which they turn back and reconstitute a new molar or molecular segment. The lines of flight used in the maps just described are of this relative kind: agents of local change within the actual world, but agents that derive their capacity to escape from a particular segmentarity from their relation to absolute lines of flight.

The idea of mapping historical processes using these three types of lines can lead to novel philosophical insights. For example, once we understand that the possibilities open to an actual assemblage have a certain virtual structure, we do not have to think about primitive societies and their urban counterparts as representing successive stages of development of humanity. Some forms of social organization may indeed have appeared earlier than others – hunter-gatherers certainly existed before any central state apparatus – but that succession occurred only in actual time. In virtual time the latter was a possibility already prefigured in the former, and it is "precisely because these processes are variables of coexistence that [they can be] the object of a *social topology* ...". [28]

In particular, primitive societies and their molecular segmentarity already contained in their associated possibility space a line of flight prefiguring a state apparatus, a line of flight that simultaneously offered an opportunity to become something else, a change of identity, as well as the risk of becoming rigidly segmented by the emergence of centralized authority. Hence, Deleuze and Guattari characterize primitive societies by the mechanisms of prevention and anticipation with which they guard off this possibility: burning all surplus food in ceremonial rituals, for example, to prevent it from becoming a reservoir of energy (a gradient) that a centralized authority can use to promote a division of labor, forcing primitives to cross the town-threshold and the state-threshold. [29]

Many other insights emerge from these metaphysical maps, although in some cases the insights are muddled by an imprecise segmentation of social reality. In particular, thinking of this reality as composed of three levels of scale, the individual, the group, and the social field, can be very misleading. We need an assemblage by assemblage break down that yields a variety of social segments in concrete part-to-whole relations: persons; communities and interpersonal networks; institutional organizations and networks or hierarchies of organizations; cities, regions and provinces; nation states, kingdoms, and empires. The description of the maps just given included this correction, the original maps being more ill defined. But considering all the philosophical labor flawlessly performed by Deleuze and Guattari, this is indeed a minor correction. Being forced to work out the details of this or that particular social segment is a small price to pay once every transcendent entity has been exorcized from an ontology, leaving behind only immanent entities as legitimate inhabitants of the material world.

REFERENCES:

1. Gordon Van Wylen. Thermodynamics. (New York: John Wiley & Sons, 1963), p. 16.

2. Gilles Deleuze. Difference and Repetition. (New York: Columbia University Press, 1994), p. 222.

3. David Greenhood. Mapping. (Chicago: University of Chicago Press,1964), Chapter 6.

4. Morris Kline. Mathematical Thought from Ancient to Modern Times. Volume 3. (New York: Oxford University Press, 1972), p. 904.

"Prior to and during the work on non-Euclidean geometry, the study of projective properties was the major geometric activity. Moreover, it was evident from the work of Von Staudt that projective geometry is logically prior to Euclidian geometry because it deals with qualitative and descriptive properties that enter into the very formation of geometrical figures and does not use the measures of line segments and angles. This fact suggested that Euclidian geometry might be some specialization of projective geometry. With the non-Euclidean geometries now at hand the possibility arose that those...might also be specializations of projective geometry."

5. Joe Rosen. Symmetry in Science. (New York: Springer-Verlag, 1995), Chapter Two.

6. Morris Kline. Mathematical Thought from Ancient to Modern Times. Op. Cit. p. 917.

7. David A. Brannan, Matthew F. Esplen, Jeremy J. Gray. Geometry. (Cambridge: Cambridge University Press, 1999). p. 364.

8. Philip Ball. Life's Matrix. A Biography of Water. (Berkeley: University of California Press, 2001), p. 161.

9. Ian Stewart and Martin Golubitsky. Fearful Symmetry. (Oxford: Blackwell, 1992), p. 108-110.

10. Gilles Deleuze and Felix Guattari. A Thousand Plateaus. (Minneapolis: University of Minnesota Press, 1987), p. 31.

"What is the significance of these indivisible distances that are ceaselessly transformed and cannot be divided or transformed without their elements changing in nature each time? Is it not the intensive character of this type of multiplicity's elements and the relations between them? Exactly like a speed or a temperature, which is not composed of other speeds or temperatures, but rather is enveloped in or envelops others, each of which marks a change in nature. The metrical principle of these multiplicities is not to be found in a homogeneous milieu but resides elsewhere, in forces at work within them, in physical phenomena inhabiting them..."

11. Ibid. p. 266.

12. Ibid. p. 6.

13. Morris Kline. Mathematical Thought from Ancient to Modern Times. Volume 3. Op. Cit. p. 882.

   "Thus if the surface of the sphere is studied as a space in itself, it has its own geometry, and even if the familiar latitude and longitude are used as the coordinates of points, the geometry of that surface is not Euclidian...However the geometry of the spherical surface is Euclidian if it is regarded as a surface in three-dimensional space." (Ibid. p. 888.) That is, the surface is not metric if its is not embedded in a global space but it becomes metric if it has a supplementary dimension from which global coordinates can be assigned.

14. June Barrow-Green. Poincare and the Three Body Problem. (Providence: American Mathematical Society, 1997), p. 32.

15. Gilles Deleuze. Difference and Repetition. Op. Cit. p. 182-183.

16. David Lewis. Counterpart Theory and Quantified Modal Logic. In The Possible and the Actual. Edited by Michael J. Loux (Ithaca: Cornell University Press, 1979), p. 117-121.

17. Gilles Deleuze. Difference and Repetition. Op. Cit. p. 208-209. (Italics in the original).

18. Gilles Deleuze and Claire Parnet. Dialogues II. (New York: Columbia University Press, 2002), p. 124-125.

19. Ibid. p. 132.

20. Gilles Deleuze and Felix Guattari. A Thousand Plateaus. Op. Cit. p. 222.

21. Ibid. p. 210-212.

22. Ibid. p. 214.

23. Ibid. p. 213.

24. Ibid. p. 215.

25. Gilles Deleuze and Felix Guattari. A Thousand Plateaus. Op. Cit. p. 251 (Italics in the original).

26. Gilles Deleuze. Logic of Sense. (New York: Columbia University Press, 1990), p. 167. (My italics).

27. Ibid. p. 162-168. I use the term "duration" in its ordinary sense, not in the technical Bergsonian sense.

28. Gilles Deleuze and Felix Guattari. A Thousand Plateaus. Op. Cit. p. 435. (My italics.)

29. Ibid. p. 433-434.

Deleuze in Phase Space.

> The semantic view of theories makes language largely irrelevant to the subject. Of course, to present a theory, we must present it in and by language. That is a trivial point... In addition, both because of our own history – the history of philosophy of science which became intensely language-oriented during the first half of [the last] century – and because of its intrinsic importance, we cannot ignore the language of science. But in a discussion of the structure of theories it can largely be ignored.
>
> Bas C. Van Fraassen. Laws and Symmetry. [1]

Van Fraassen is perhaps the most important representative of the empiricist tradition in contemporary analytical philosophy. But why use a quote from an analytical philosopher, however famous, to begin a discussion of the work of an author who many regard as a member of the rival continental school of philosophy?. The answer is that Gilles Deleuze does not belong to that school, at least if the latter is defined not geographically but in terms of its dominant traditions (Kantian and Hegelian). As is well known, Deleuze himself argued for the superiority, in some respects, of anglo-american, or empiricist, philosophy, from Hume to Russell. [2] In addition, Deleuze's work was in large part a sustained critique of language (or more generally, of representation) as the master key to philosophical thought and, as the opening quote attests, Van Fraassen is also a leader of the emerging faction of philosophers of science disillusioned with the linguistic approach. There are, then, points of convergence between the two authors, but there are also also several divergences. This essay will explore both.

Let's first of all clarify Van Fraassen's position. What does it mean to say that in discussing the structure of scientific theories the language in which they are expressed is irrelevant? Or to put it differently, in what approach towards the nature of scientific theories is language itself crucial, and why is that approach, according to Van Fraassen, wrong? The approach in question is the *axiomatic approach* to science according to which the content of a theory may be modeled by a set of axioms, or self-evident truths, and all the theorems that may be derived from the axioms using deductive logic. Although there are many versions of this approach – some regarding the axioms purely syntactically, others treating them as part of natural language – what they all have in common is a disregard for the actual mathematical tools used by scientists, tools like the differential calculus. It is through the use of nonlinguistic tools that scientists create models of physical phenomena, and it is these models that have become the object of intense interest for analytical philosophers. The question of whether the set of models that makes up a theory is axiomatizable or not, that is, whether they can be given a homogenous hierarchical logical structure or not, is still a valid question but has now become less important since, for all we know, a theory's models may constitute a *heterogeneous population* accumulated over time. [3]

To make things worse for the axiomatic approach the heterogeneity of this population has increased in the last one hundred years. While before most models used differential equations as their basis, suggesting that there may exist a general theory of models, in the last century many other kinds of equations (finite difference equations, matrix equations) have been added to the modeling resources available to scientists. [4] More recently digital computers have increased this diversity with modeling tools like cellular automata, Monte Carlo simulations, genetic algorithms, neural nets. In this essay I will explore mostly the oldest modeling technology, the one based on the differential calculus, partly because it is the one better understood, and partly because it is the one actually discussed by both Deleuze and Van Fraassen.

To use differential equations as a model one must first specify all the relevant ways in which a physical system is free to change, that is, its "degrees of freedom". As the degrees of freedom of a system change its overall state changes. This implies that a model of the system must capture the different possible states in which it can exist. This set of states may be represented as a *space of possibilities* with as many dimensions as the system has degrees of freedom. This space is referred to as "state space" or "phase space". In this space each point represents one possible state for a physical system, the state it has at a given instant of time. As the states of a physical system change with time, that is, as the system goes through a temporal sequence of states, its representation in state space becomes a continuous sequence of points: a curve or a trajectory. [5] Each point in this space, each possible state, may have different probabilities of existing. A space in which all the points are equally probable is a space without any structure, and it represents a physical system in which states change in a completely random way (an ergodic system).

Van Fraassen discusses two ways in which the possibility space can be given structure: through rules that restrict the areas that may be occupied – thus assigning different probabilities to different parts of the space, including forbidden areas with zero probability – and through rules that specify what states must follow other states, that is, through rules governing trajectories. Van Fraassen refers to these two kinds of rules as "laws of coexistence", exemplified by Boyle's law of ideal gases, and "laws of succession", exemplified by Newton's laws of motion. [6] Both types of rules are expressed as equations, so for Van Fraassen it is the equations that give us the structure of the space of possibilities. Deleuze, as I will argue shortly, gives a more original account of this structure, one that does not depend on the concept of "law". But the main point remains the same in both accounts: if all possible states are equiprobable, then no regularity may be discerned in the dynamics of a system, so whatever breaks the equiprobability is what is philosophically interesting. Before describing the Deleuzian solution to this problem, or rather, the way in which Deleuze poses the problem, we need to discuss in more detail how mathematical models work.

Any equation, whether differential or not, has numerical solutions, that is, sets of values for its unknown variables that make the equation come out true. Each numerical solution represents one state of the system being modeled. But in order to learn about a physical system scientists need to know more than just a few numerical solutions: they must have a sense of *the pattern formed by all numerical solutions* of a given equation.

When this global pattern is given by yet another equation, it is called an "exact" or a "analytical" solution. Without being exactly solvable an equation is of limited value as a model because it gives us no information about the overall pattern, and without that we cannot generalize from particular cases. Historically, the main incentive behind the development of phase space was the resistance that some recalcitrant equations offered to being exactly solved: *nonlinear* differential equations in which there are interactions between the degrees of freedom. Because of the impossibility to obtain analytical solutions for many nonlinear equations classical physicists used mostly linear models, models capturing only the simplest behavior of a material system. And in the few cases when nonlinear equations were used they were restricted to *low intensity* values of the variables, a range of values for which their behavior was effectively linearized. This, at best, limited the kinds of physical phenomena that could be modeled, and at worse, led to the false idea that the world is in fact linear, like a giant clockwork mechanism. As the mathematician Ian Stewart puts it:

> Classical mathematics concentrated on linear equations for a sound pragmatic reason: it could not solve anything else... So docile are linear equations, that classical mathematicians were willing to compromise their physics to get them. So the classical theory deals with *shallow* waves, *low*-amplitude vibrations, *small* temperature gradients. So ingrained became the linear habit that by the 1940's and 1950's many scientists and engineers knew little else. ... Linearity is a trap. The behavior of linear equations ... is far from typical. But if you decide that only linear equations are worth thinking about, self-censorship sets in. Your textbooks fill with triumphs of linear analysis, its failures buried so deep that the graves go unmarked and the existence of the graves goes unremarked. As the eighteenth century believed in a clockwork world, so did the mid-twentieth in a linear one. [7]

Phase space was created to overcome these limitations. By studying the possibility spaces defined by the calculus, instead of the differential equations themselves, mathematicians could forget about exact solutions and concentrate on the question of whether the spaces had any structure that broke the equiprobability of the states, a structure defined by special, or singular, points. One thing that made these points special was that they remained unaltered if the space was transformed in a variety of ways. That is, these singularities constituted the most stable and characteristic aspect of the space. In addition, when the physical system being modeled was not isolated so that it could dissipate heat to its surroundings, a singularity acted as an *attractor*, that is, it was capable of affecting nearby trajectories forcing then to converge on it. Thus, given that the set of trajectories is the geometrical counterpart of the numerical solutions to the equation, and that their overall behavior in phase space is governed by these special points, *the distribution of singularities* gives us information about the pattern of all the solutions. This is not the same as having an analytical solution, but it is the next best thing. [8]

This is, in a nut shell, the reason why phase space commands so much attention today: as philosophers of science have turned away from the linguistic to the mathematical expression of scientific concepts the study of differential equations (and of the behavior of their solutions) has become top priority, and the main way to study these equations if they are nonlinear is using the geometrical approach. But this still does not explain in what sense phase space is important outside the philosophy of science, that is, what insights can such spaces yield about the material reality studied by scientists. If these models were nothing but mathematical constructs there would be no reason to think they may throw some light on the nature of reality. But many of these models actually work, that is, they manage to capture the regularities in the behavior of real systems.

Let's assume that we have a laboratory where we can manipulate real physical systems, that is, where we can restrict their degrees of freedom (by screening out other factors) and where we can place a system in a given state and then let it run spontaneously through a sequence of states. Let's also assume that we can measure with some precision the values of the degrees of freedom (say, temperature, pressure and volume) at each of those states. After several trials we generate data about the system starting it at different initial states. The data will consist, basically, of sequences of numbers giving the values of temperature, pressure and volume that the system takes as it evolves from different initial conditions. We can plot these

number series in a piece of paper turning them into a curve or
trajectory. We then run our mathematical model, giving it the
same values for initial conditions as our laboratory runs, and
generate a set of phase space trajectories. Finally, we compare
the two sets of curves. If the mathematical and experimental
trajectories display geometrical similarity this will be evidence
that the model actually works. As one analytical philosopher
puts it:

> we can say that a dynamical theory is approximately true just
> if the modeling geometric structure approximates (in suitable respects)
> to the structure to be modeled: a basic case is where trajectories in the
> model closely *track trajectories* encoding physically real behaviors (or,
> at least, track them for long enough). [9]

It is only when mathematical models have the capacity
to track the results of laboratory experiments that there is a
philosophical justification to perform an ontological analysis of
phase space. This analysis is needed because the tracking ability
of models must be given an explanation, unless we are prepared
to accept it as a brute fact, or worse yet, as a unexplainable
miracle. Assessing the ontological status of phase space, on the
other hand, necessarily goes beyond both mathematical
representations and laboratory interventions, involving the most
basic metaphysical presuppositions.

One may presuppose, for example, the autonomous
existence of objects of direct experience (pets, automobiles,
buildings) but assume that entities like oxygen, electrons, or
causal relations, are mere theoretical constructs. Presuppositions
of this sort are associated with positivism and empiricism,
though different philosophers will draw the line of what is
"directly observable" at different places. Van Fraassen, for
instance, seems to believe that objects perceived through
telescopes, but not microscopes, count as directly experienced. [10]
Realist philosophers, on the other hand, tend to reject the
distinction between the observable and the unobservable as too
anthropocentric, although they too may differ on what they
believe are the contents of a mind-independent world. Deleuze is
a realist philosopher, but one determined to populate an
autonomous reality exclusively with immanent entities, and to
exorcise from it any transcendent ones, like Aristotelian

essences. Thus, the first point of divergence between Deleuze and Van Fraassen is one of different *ontological commitments*.

The first candidates for ontological evaluation in the case of phase space are the trajectories themselves. These, as I said, represent possible sequences of states. Empiricists are notoriously skeptical about possible entities. Quine, one of the most famous representatives of this school, is well known for the fun he pokes at these entities. As he writes: "Take, for instance, the possible fat man in the doorway; and again, the possible bald man in the doorway. Are they the same possible man, or two possible men? How do we decide? How many possible men there are in that doorway? Are there more possible thin ones than fat ones? How many of them are alike? Or would their being alike make them one?." [11] In other words, Quine is arguing that we do not have the means to *individuate* possible entities, that is, to identify them in the midst of all the possible variations. There is simply not enough structure in a possible world to know whether we are dealing with one or several entities as we modify the details. But, it may be argued, this is a problem only for *linguistically* specified possible worlds.

The target of Quine's ridicule is the modal logician who believes that the fact that people can understand counterfactual sentences, sentences like "If J.F.K. had not been assassinated the Vietnam War would have ended sooner.", implies the objective existence of possible worlds. But as realist philosophers like Ronald Giere have argued, while Quine's skeptical remarks are valid for counterfactuals, the extra structure that phase space possesses can overcome these limitations:

As Quine delights in pointing out, it is often difficult to individuate possibilities. ... [But] many models in which the system laws are expressed as differential equations provide an unambiguous criterion to individuate the possible histories of the model. They are the trajectories in state space corresponding to all possible initial conditions. Threatened ambiguities in the set of possible initial conditions can be eliminated by explicitly restricting the set in the definition of the theoretical model. [12]

Let's assume for a moment that Giere is right and that, within the restricted world of phase space, the possible histories of a system can be individuated. Van Fraassen could still deny the need for an ontological commitment to modalities given that for him the point of building theoretical models is simply to achieve *empirical adequacy*, that is, to increase our ability to make predictions and to control outcomes in the laboratory. For this purpose all that matters is that we generate a single trajectory for a given initial condition; reproduce that particular combination of values for the degrees of freedom in the laboratory; and observe whether the sequence of *actual states* matches that predicted by the trajectory. Given the single phase space trajectory that we associate with the actual sequence of states in an experiment, the rest of the population of trajectories is merely a useful fiction, that is, it is ontologically unimportant. Giere refers to this ontological stance towards modalities as "actualism". [13]

But as Giere goes on to argue, this ontological stance misses the fact that the population of trajectories as a whole displays certain regularities in the possible histories of a system, global regularities that play a role in shaping any one particular actual history. In the terms used above, the space of possibilities has structure, and this structure is not displayed by any one single trajectory. For Giere understanding a system is not just knowing how it actually behaves in this or that specific situation, but knowing *how it would behave* in conditions that may not in fact occur. And to know that we need to use the global information embodied in the population of possible histories.

Van Fraassen may reply, of course, that this information is given by the laws of succession that control the evolution of trajectories. This would seem to commit him, however, to assert the existence of another modal property, necessity, since when laws are considered to be objective features of reality they are typically assumed to be necessary features. The problem here is that necessity and possibility are interdefinable modal concepts: if an event must necessarily occur then it is not possible that it would not occur. But when empiricists or positivists speak of laws they do not usually refer to the objective regularities exhibited by material processes, regularities that may not always be directly observable, but to the equations capturing those regularities, equations being directly observable when printed on a piece of paper. Thus, the debate between realists and empiricists seems to offer only two alternatives: either be ontologically committed to traditional modalities or reject the latter but loose your ability to explain why there are recurrent regularities in the world. This alternative, however, is a trap, and the significance of Deleuze's realist approach is precisely that it supplies us with a escape route.

Deleuze is not an actualist but he is not a realist about traditional modalities either. Rather, he invents a new form of physical modality to account for both the regularities in the models and the immanent patterns of becoming in nature. This new modality he refers to as "virtuality", *the ontological status of something that is real but not actual*. The tendencies and causal capacities of material entities, when not actually manifested or exercised, are virtual in this sense. But how is this different from saying that an unmanifested tendency, or an unexercised capacity, is merely possible? Because what possesses this virtual status is not the space of possibilities defined by a tendency or capacity, but the structure of such a space. I said above that this structure is given by a distribution of singularities, breaking the equiprobability of possible states. Singularities, on the other hand, can help us conceptualize the modal status of tendencies, not of capacities, so I will restrict my remarks to the former.

First of all, we need to bring into the discussion another component of phase space: *the velocity vector field*. Because Poincaré was investigating the structure of the space of possible numerical solutions to differential equations, the abstract spaces used were differential manifolds, not metric Euclidean spaces. While a metric space is a set of points defined by global coordinates, in a differential manifold the component points are defined using only local information: the instantaneous rate of change of curvature at a point. This means that rather than being a set X, Y, and Z coordinates, a differential manifold is *a field of rapidities and slownesses*: the rapidity or slowness with which curvature changes at each point. In fact, every point is not only a speed but a velocity, since a direction may be assigned to it. A velocity can be represented by a vector, so the entire space possesses a field of velocity vectors. The importance of this is that the form that the integral curves or trajectories take in phase space is determined by the vector field.

Hence, while Giere's insight that there is information in the population of curves is entirely correct, he does not seem to realize that that extra information reflects properties of the vector field, like the property of possessing a certain distribution of singularities. While the nature of a singularity must be established through the use of nearby trajectories – whether a point singularity is a focus or a node, for example, is determined by observing how the integral curves in its vicinity approach it, spirally or on a straight line – the *existence and distribution* of the singularities does not need any trajectories to be established. As Deleuze writes:

> Already Leibniz had shown that the calculus...expressed problems which could not hitherto be solved or, indeed, even posed... One thinks in particular of the role of the regular and the singular points which enter into the complete determination of the species of a curve. No doubt the specification of the singular points (for example, dips, nodes, focal points, centers) is undertaken by means of the form of integral curves, which refers back to the solutions of the differential equations. There is nevertheless a complete determination with respect to the existence and distribution of these points which depends upon a completely different instance, namely, the field of vectors defined by the equation itself.... Moreover, if the specification of the points already shows the necessary immanence of the problem in the solution, its involvement in the solution which covers it, along with the existence and distribution of points, testifies to the transcendence of the problem and its directive role in relation to the organization of the solutions themselves". [14]

Thus, the first step in this alternative interpretation consists in sharply differentiating these two components of phase space, the population of trajectories and the vector field, a step that, to my knowledge, has not been taken by any analytical philosopher. It may be objected that adding vector fields to the list of things that must be given an ontological interpretation would bring us back to the endless and fruitless discussions that, from the time of Leibniz to the early nineteenth century, tended to surround the notion of an "infinitesimal quantity", because each vector in the field is one such infinitesimal. These entities were eliminated from the foundations of the calculus by the concept of a limit, a concept that presupposes only the notion of number and nothing else. But what must be given an ontological interpretation is not the vectors themselves but *the topological invariants of the entire field*, and these have nothing whatsoever to do with infinitesimals.

A clue to the modal status of these invariants is the fact that trajectories always approach an attractor *asymptotically*, that is, they approach it *indefinitely closely but never reach it.* [15] Although the sphere of influence of an attractor, it's basin of attraction, is a subset of points of phase space, and therefore a set of possible states, the attractor itself is not a possible state since it can never become actual. In other words, unlike trajectories representing possible histories that may or may not be actualized, attractors can never be actualized since no point of a trajectory can ever reach them. Despite their lack of actuality attractors are nevertheless real since they have definite effects. In particular, they confer on trajectories a strong form of stability, called "asymptotic stability". [16] Small shocks may dislodge a trajectory from its attractor but as long as the shock is not too large to push it out of the basin of attraction, the trajectory will spontaneously return to the stable state defined by the attractor. It is in this sense that singularities represent only the long term tendencies of a system but never a possible state. Thus, it seems, that we need a new form of physical modality, distinct from possibility and necessity, to account for this double status of singularities: real in their effects but incapable of ever being actual. This is what the notion of virtuality is supposed to achieve.

The second point of divergence between Van Fraasen and Deleuze can be summarized as follows. Both philosophers agree that the value of phase space is that it gives us a means to study the content of scientific theories that cannot be expressed by propositions, that is, by the meaning of declarative sentences. But for Van Fraasen the only important component of phase space is the individual trajectory, a single *solution* picked for the purpose of using it to predict the outcome of a laboratory experiment. For Deleuze, on the other hand, the components that matter are those that allow us to capture the extra-propositional and sub-representative nature not of solutions, but of scientific

and philosophical *problems.* [17] A problem can be defined entirely
independently of its solutions, by *a distribution of the significant
and the insignificant,* and within the significant, by a distribution
of the singular and the ordinary. [18]

Posing a problem in classical mechanics, for example,
involves first of all discerning the significant ways in which a
process can change, it's degrees of freedom, and discarding all
other ways of changing as trivial or insignificant. A simple
system like a pendulum, for example, can change in only two
ways, position and speed. We could, of course, explode the
pendulum or melt it at high temperatures, and these would also
constitute ways in which it can change, they would just not be
relevant ways of changing from the point of view of its intrinsic
dynamics. Thus, creating the phase space of a pendulum, posing
the dynamical pendulum problem, implies an assessment of
significance, since the relevant degrees of freedom become the
dimensions of the possibility space. Then, having determined
one topological invariant (the number of dimensions) we explore
the vector field to find what other invariants additionally specify
the conditions of the problem: a distribution of singularities.
Since all this can be achieved without having to think about
solutions (trajectories) it is clear that a problem must be thought
as possessing an objectivity independently of its solutions. A
problem ceases to be a transitory subjective state of ignorance of
a solution and becomes an objective entity that does not cease to
exist once it is solved.

Moreover, for Deleuze not only are problems
independent of their solutions, they have a genetic relation with
them: *a problem engenders its own solutions as its conditions
become progressively better specified.* Deleuze's discussion of
this point uses a different branch of mathematics, group theory,
and its application to the solutions of algebraic, not differential,
equations, so we first need to give the historical background of
this other field. As mentioned above, there are two kinds of
solutions to equations, numerical and analytical. A numerical
solution is given by numbers that, when used to replace an
equation's unknowns, make the equation come out true. For
example, an algebraic equation like $x2 + 3x - 4 = 0$ has as its
numerical solution $x = 1$. An analytical or exact solution, on the

other hand, does not yield any specific value or set of values but rather the global pattern of all numerical solutions, a pattern expressed by another equation or formula. The above example, which may be re-written as $x^2 + ax - b = 0$, has the analytical solution: $x = \sqrt{(a/2)^2 + b} - a/2$. By the sixteenth century mathematicians knew the exact solutions to algebraic equations where the unknown variable was raised up to the fourth power (that is, those including x^2, x^3 and x^4). But then a crisis ensued. Equations raised to the fifth power refused to yield to the previously successful method.

The breakthrough came two centuries later when it was noticed that there was a pattern to the solutions of the first four cases, a pattern that might hold the key to understand the recalcitrance of the fifth, known as *the quintic*. The mathematicians Neils Abel and Evariste Galois found a way to approach the study of this pattern using resources that today we recognize belong to group theory. In a nut shell we can say that Galois "showed that equations that can be solved by a formula must have groups of a particular type, and that the quintic had the wrong sort of group." [19] The term "group" refers to a set of entities with special properties, and a rule of combination for those entities. The most important of the properties is the one referred to as "closure", which means that when we use the rule to combine any two entities in the set, the result is also an entity in the set. The set of positive integers, for example, forms a group under the rule of addition, but not under subtraction, because the latter can yield negative integers, that is, elements not in the set. Another example is the set of rational numbers that forms a group under multiplication, but not under division (which may yield irrational numbers). For our purposes here the most important groups are those whose members are *transformations*, and the rule a consecutive application of those transformations. For example, the set consisting of rotations by ninety degrees (that is a set containing rotations by 90, 180, 270, and 360 degrees) forms a group, since any two consecutive rotations produce a rotation also in the group.

To understand how a group of transformations can be used to establish the conditions of a problem, that is, to generate a distribution of the significant and the insignificant, let's give a concrete example: the use of groups of transformations to study the invariants of physical laws. For the laws of classical physics, the group includes displacements in space and time, as well as rotations and other transformations. Let's we imagine a physical phenomenon that can be reliably produced in a laboratory. If we displace it in space – by reproducing the phenomenon in another, far away laboratory – we will leave all its properties invariant. Similarly, if we simply change the time at which we begin an experiment, we can expect this time displacement to be irrelevant as far as the regularity of the phenomenon is concerned. It is only the difference in time between the first and final states of the experiment that matters, not the absolute time at which the first state occurs. Thus, via transformations applied to the equations expressing laws we can discover those types of change to which *the law is indifferent*, or the type of changes that do not make a difference to it, and conclude that using absolute time or absolute position as inputs to the equation expressing a law is irrelevant.

In a similar way, Galois used certain transformations (*substitutions or permutations* of the solutions) that, as a group, revealed the invariances in the relations between solutions. More specifically, when a permutation of one solution by another left the equation valid, the two solutions became *indistinguishable* as far as their validity was concerned. The group of an equation became a key to its solvability because it expressed the degree of indistinguishability of the solutions. [20] Or as Deleuze would put it, the group revealed not what we know about the solutions, but *the objectivity of what we do not know about them*, that is, the objectivity of the problem itself. [21] And Deleuze goes on to emphasize that, beside demonstrating the autonomy of problems from solutions, the group theoretic approach to the quintic shows that the solutions to the equation are produced as the original group gives rise to subgroups that successively limit the substitutions that leave relations invariant. That is, the problem gives birth to its solutions as its own conditions become progressively better defined. As he puts it:

We cannot suppose that, from a technical point of view, differential calculus is the only mathematical expression of problems as such. ... More recently other procedures have fulfilled this role better.

Recall the circle in which the theory of problems was caught: a problem is solvable only to the extent that is is 'true' but we always tend to define the truth of a problem by its solvability. ... The mathematician Abel was perhaps the first to break this circle: he elaborated a whole method according to which solvability must follow from the form of a problem. Instead of seeking to find out by trial and error whether a given equation is solvable in general we must determine the conditions of the problem which progressively specify the fields of solvability in such a way that the statement contains the seed of the solution. This is a radical reversal of the problem-solution relation, a more considerable revolution than the Copernican. ... The same judgement is confirmed in relation to the work of Galois: starting from a basic 'field' (R), successive adjunctions to this field (R', R'', R'''...) allow a progressively more precise distinction between the roots of an equation by the progressive limitation of possible substitutions. There is thus a succession of 'partial resolvents' or an 'embedding of groups', which make the solutions follow from the very conditions of the problem. [22]

Thus, for Deleuze there can be several ways of using mathematics to express problems as such, all of which are part of the non-linguistic content of scientific fields. This is an important insight for the philosophy of science. But as argued above, the consequences of these ideas that go beyond the world of science, the consequences for a realist ontology, demand that we connect the ideas to the material world, at least in the controlled setting of laboratory experiments. We need two establish two connections: one for the idea that problems have an existence separate from their solutions, and another for the idea that the gradual definition of a problem's conditions is involved in the production of solutions.

The autonomy of mathematical problems from their solutions suggests that physical, chemical, biological and other problems also exist virtually, independently of any actual solution. Let's give the simplest example of an objective problem: *an optimization problem*. Many physical entities, like bubbles or crystals, must solve an optimization problem as they constitute themselves: a bubble must find the shape that minimizes surface tension, while the crystal must find the shape that minimizes bonding energy. If we created a phase space representation of the dynamics that lead to the spherical shape of bubbles, or the cubic shape of salt crystals, we would find a single point singularity structuring the space, a topological point representing a minimum of something. What exactly is minimized varies from case to case, but the structure of the possibility space is the same. It is clear, however, that once the spherical or cubic shapes have emerged as solutions, the optimization (or minimization) problem does not disappear, since all bubbles and salt crystals in the future will still have to solve it as they form.

The second connection can be established by combining the resources of group theory and of dynamical systems theory, as the geometrical approach to the study of differential equations is known. There are mathematical events, known as *bifurcations*, that can transform one distribution of attractors into a topologically inequivalent one. Sometimes it is only the number of attractors that changes, but other times it is their type: a point attractor can be changed into a line attractor shaped as a loop by an event called a *Hopf bifurcation*; and the resulting loop (a periodic attractor) can be changed into a chaotic attractor by an event called a *Feigenbaum bifurcation*. These mathematical events result from operations applied to the vector field of a phase space: a small vector field is added to the main one to perturb it, and when the perturbation reaches a critical threshold, a bifurcation results. [23] A sequence of such events has a similar group theoretic structure as the series of permutations used by Galois to generate the solutions to the quintic equation, a structure sometimes referred as a "symmetry-breaking cascade". That is, as the bifurcations transform one singularity (or set of singularities) into another, all the solutions to the problem posed by the differential equations progressively unfold: steady-state solutions, periodic solutions, chaotic solutions.

Such a cascade of broken symmetries can also be found in the material world. A moving fluid, for example, must solve the problem of how to flow at different speeds. At at slow speeds the solution is simple: stick to steady-state or uniform flow. But after a critical threshold of speed is crossed that solution becomes insufficient and the moving fluid must switch to a convective or wavy flow. Finally, after another critical threshold, the faster speeds pose a flow problem to the fluid that it cannot solve by moving rhythmically and it is forced to become turbulent. Although in controlled laboratory experiments we can discover a larger number of regimes of flow, these three will suffice for present purposes. [24] The basic idea is that, just like trajectories in phase space can track series of measurement of the state of a physical system, a sequence of mathematical events can track a sequence of physical events: phase transitions between one regime of flow and another. The main difference between the two cases is that while trajectories track plotted measurements *quantitatively*, and the relation between a trajectory and a plot is one of geometrical similarity, bifurcations track phase transitions *qualitatively*, and there is no geometrical resemblance between the singularities and the regimes of flow.

To say that the tracking is only qualitative is not to imply that it is any less serious. Rather, as Deleuze would put it, it is "anexact yet rigorous". [25] This follows from the very nature of topology or differential geometry, the geometries used to create phase space, in which the concepts of exact length, area, or volume are meaningless, but in which comparison of other properties can be done in a rigorous way. Furthermore, Deleuze would argue that the qualitative is more fundamental than the quantitative in the sense that the former can generate the latter as a special case. It is well known that a continuous and qualitative topological space can generate a discontinuous and quantitative metric space as the groups of transformations associated with the former get progressively smaller, and as the geometric spaces become progressively more discontinuous and quantitative, following the sequence: topology, differential geometry, projective geometry, affine geometry, and finally metric (Euclidean and non-Euclidean) geometry. [26]

This full symmetry-breaking cascade is not present in the case of phase space, but it is nevertheless clear that the topological invariants, the singularities in the vector field, are genetically prior to the metric trajectories, or put differently, the qualitatively specified problem is genetically more fundamental than the quantitatively defined solutions. In a previous quote Deleuze compared the achievements of Abel and Galois to the revolutionary impact of astronomy's switch to heliocentrism. But, as he adds elsewhere, the gains from group theory can only be realized philosophically if we blend its insights with those of non-metric geometries to create a theory of problems:

> Solvability must depend upon an internal characteristic: it must be determined by the conditions of the problem, engendered in and by the problem along with the real solutions. Without this reversal, the famous Copernican revolution amounts to nothing. Moreover, there is no revolution so long as we remain tied to Euclidean geometry: we must move to a geometry of sufficient reason, a Riemannian-like differential geometry which *tends to give rise to discontinuity on the basis of continuity*, or to ground solutions in the conditions of the problem. [27]

The priority of the qualitative over the quantitative is not confined to mathematical models. In Deleuze the genetic link between the two is first and foremost constitutive of material reality. To return to a previous example, a topological point yields two different metric shapes, spheres and cubes, as it differentiates, or what amounts to the same thing, a qualitatively specified virtual problem produces two quantitative physical solutions as it is *divergently actualized*. Indeed, the virtual problem is not only actualized into soap bubbles, salt crystals, and other physical forms that emerge as solutions to an optimization problem. It is also actualized in mathematical models sharing the same topological invariants. Hence, the epistemological relation between a mathematical model and the process that it models is one of *co-actualization* of the same virtual problem, a relation that would explain why mathematical models work at all. [28]

The divergent actualization of the virtual structure of possibility spaces is also important to Deleuze as evidence that he has not reintroduced Platonic essences into his ontology, since essences like "sphericity" resemble that which incarnates them, whereas there is no resemblance between topological and metric entities. But are we not introducing transcendent entities through the back door when we speak of "invariants", topological or otherwise?. If the term "invariant" was used by itself, it would certainly carry implications of an eternally unchanging essence. But in its technical meaning it is always used *relative to a transformation*: rotations, displacements, projections, stretchings, foldings. And as Deleuze argues these transformations refer to the capacities to affect of mathematical operators, as well as to the capacity to be affected of the data structures (numbers, matrices, equations) these operators affect. The application of an operator to its target, the addition of a small vector field to the main one causing a bifurcation to occur, for example, is always an event. And events, unlike essences, are not necessary but contingent. Unlike what happens in an axiomatic approach, in a problematic conception of geometry:

> ... figures are considered only from the view point of the *affections* that befall them: sections, ablations, adjunctions, projections. One does not go by specific differences from a genus to its species, or by deduction from a stable essence to the properties deriving from it, but rather from a problem to the accidents that condition it and resolve it. This involves all kinds of deformations, transmutations, passages to the limit, operations in which each figure designates an 'event' much more than an essence; the square no longer exists independently from a quadrature, the cube from a cubature, the straight line from a rectification. Whereas the theorem belongs to the rational order, the problem is affective and is inseparable from the metamorphoses, generations, and creations within science itself. [29]

To conclude: despite Van Fraasen's avowed goal to avoid a linguistic approach to science, Deleuze would argue that his ontological commitments place him in the same tradition that, from Aristotle to Kant, have traced problems from the propositions that express cases of solutions. It does not really matter how the solutions are expressed, linguistically or non-linguistically, whether as "logical opinions, geometrical theorems, algebraic equations, physical hypotheses, or transcendental judgements." [30] What matters is the subordination of problems to solutions, a subordination that threatens to negate the gains from the Copernican revolution by focusing the efforts of scientists and philosophers of science on final products (physical, chemical, biological solutions) instead of on the processes that produce those products, as virtual problems become progressively better specified. It is only by inverting the traditional relation between problems and solutions, and by systematically pursuing all the philosophical consequences of this radical inversion, that we can free science and philosophy from the straight jacket of axioms and theorems.

REFERENCES:

1. Bas Van Fraassen. Laws and Symmetry. (Oxford: Clarendon Press, 1989), p. 222.

2. Gilles Deleuze and Claire Parnet. Dialogues II. (New York: Columbia University Press, 2002), Chapter 2.

3. Ronald N. Giere. Explaining Science. A Cognitive Approach. (Chicago: The University of Chicago Press, 1988), p. 82.

4. Mario Bunge. Causality and Modern Science. (New York: Dover, 1979), p. 75.

5. Ralph Abraham and Christopher Shaw. Dynamics: the Geometry of Behavior. Volume One. (Santa Cruz: Aerial Press,1985), p. 20-21.

6. Bas Van Fraasen. Laws and Symmetry. Op. Cit. p. 223.

7. Ian Stewart. Does God Play Dice?. The Mathematics of Chaos. (London: Basil Blackwell, 1989), p. 83.

8. June Barrow-Green. Poincare and the Three Body Problem. (Providence: American Mathematical Society, 1997). p. 32-38.

9. Peter Smith. Explaining Chaos. (Cambridge: Cambridge University Press, 1998), p. 72. (My italics.)

10. Bas Van Fraasen. The Scientific Image. (Oxford: Clarendon Press, 1980), p. 16.

11. Willard Van Orman Quine. Quoted in Nicholas Rescher. The Ontology of the Possible. In The Possible and the Actual. Michael J. Loux ed. (Ithaca: Cornell University Press, 1979), p. 177.

12. Ronald N. Giere. Constructive Realism. In Images of Science. Edited by Paul M. Churchland and Clifford A. Hooker. (Chicago: The University of Chicago Press, 1985), p. 43-44.

13. Ibid. p. 44.

14. Gilles Deleuze. Difference and Repetition. (New York: Columbia University Press, 1994), p. 177.

15. Ralph Abraham and Christopher Shaw. Dynamics: the Geometry of Behavior. Volume Three. Op. Cit. p. 35-36.

16. Gregoire Nicolis and Ilya Prigogine. Exploring Complexity. (New York: W.H. Freeman, 1989), p. 65-71.

17. Gilles Deleuze. Difference and Repetition. Op. Cit. p. 178.

18. Ibid. p. 189.

19. Ian Stewart and Martin Golubitsky. Fearful Symmetry. (Oxford: Blackwell, 1992), p. 42. (Italics in the original.)

20. Morris Kline. Mathematical Thought from Ancient to Modern Times. Volume 2 (New York: Oxford University Press, 1972), p. 759.

21. Gilles Deleuze. Difference and Repetition. Op. Cit. p. 162.

22. Ibid. p. 179-180.

23. Ralph Abraham and Christopher Shaw. Dynamics: the Geometry of Behavior. Op. Cit. Volume Three. p. 37-41.

24. Ian Stewart and Martin Golubitsky. Fearful Symmetry. Op. Cit. p. 108-110.

25. Gilles Deleuze and Felix Guattari. A Thousand Plateaus. (Minneapolis: University of Minnesota Press, 1987), p. 483.

26. Lawrence Sklar. Space, Time, and Space-Time. (Berkeley: University of California Press, 1977), p. 49-54.

27. Gilles Deleuze. Difference and Repetition. Op. Cit. p. 162.

28. Manuel DeLanda. Intensive Science and Virtual Philosophy. (London: Continuum, 2002.), Chapter Four.

29. Gilles Deleuze and Felix Guattari. A Thousand Plateaus. Op. Cit. p. 362. (Italics in the original.) See also:

Gilles Deleuze. Difference and Repetition. Op. Cit. p. 187-189.

30. Ibid. p. 161.

CPSIA information can be obtained
at www.ICGtesting.com
Printed in the USA
BVHW030900261218
536402BV00002B/201/P